H. Bird

D1539265

Illustrated Guide to Trees and Shrubs

Illustrated Guide to Trees and Shrubs

A HANDBOOK OF THE WOODY PLANTS
OF THE NORTHEASTERN UNITED STATES
AND ADJACENT CANADA

REVISED EDITION

ARTHUR HARMOUNT GRAVES

DOVER PUBLICATIONS, INC.
New York

Published in Canada by General Publishing Company, Ltd., 30 Lesmill
Road, Don Mills, Toronto, Ontario.
Published in the United Kingdom by Constable and Company, Ltd., 3
The Lanchesters, 162–164 Fulham Palace Road, London W6 9ER.

This Dover edition, first published in 1992, is an unabridged, unaltered
republication of the Revised Edition (1956) of the work first published by
Harper & Brothers, New York, 1952, under the title *Illustrated Guide to
Trees and Shrubs: A Handbook of the Woody Plants of the Northeastern
United States and Adjacent Regions.*

Manufactured in the United States of America
Dover Publications, Inc., 31 East 2nd Street, Mineola, N.Y. 11501

Library of Congress Cataloging in Publication Data

Graves, Arthur Harmount, 1879–
    Illustrated guide to trees and shrubs : a handbook of the woody plants
of the northeastern United States and adjacent Canada / Arthur
Harmount Graves.—Rev. ed.
        p.   cm.
    Originally published: Rev. ed. New York : Harper, [1956].
    Includes bibliographical references and index.
    ISBN 0-486-27258-3 (pbk.)
    1. Woody plants—Northeastern States—Identification.   2. Woody
plants—Canada, Eastern—Identification.   I. Title.
QK118.G73   1992
582.1'50974—dc20   92-9525
                                                                         CIP

TO MY PUPILS

*who have roamed with me through woods
and fields—and by their enthusiastic
and earnest queries
have kept me alive to the great variability
and the great constancy of Nature*

PREFACE

This book is intended to serve as a guide to all woody plants — trees, shrubs, or vines — whether wild or cultivated, commonly found in the Northeastern United States. By "cultivated" plants I mean those that have been planted, whether native, naturalized or exotic. If your plant has a *woody* stem, is hardy, i. e., will endure our winter climate, and is of fairly common occurrence, this book has been planned to help you to identify it. Especially if one has acquired a new piece of property, or perhaps has long owned one, with the plant life of which he wants to become acquainted, this book should enable him to identify any woody plants growing thereon.

Because this is the primary purpose of the book I have tried to keep it small so that it can be taken into the field, and perhaps carried in the pocket. Therefore, I have omitted lengthy technical descriptions: these can be found in the manuals. Following the method in a "Teaching Guide to Trees and Shrubs of Greater New York"* by Hester M. Rusk and myself, only the most important features of each plant are pointed out. Also, details of flower structure have generally been omitted, although, since the natural classification depends primarily on these, in some cases, as in the Rose Family, reference to them has been unavoidable.

The book is in large part an illustrated edition of the "Teaching Guide" above mentioned, but covering a larger area and including the common cultivated woody plants. It has been prepared because of the repeated urging of many of the pupils in my classes, held at the Brooklyn Botanic Garden and elsewhere during the past forty years.

I suggest that before trying out the keys for identification the "non-botanist" read over carefully the introduction and the section on "How to use the keys." For example, he will need to understand the difference between simple and compound leaves and between a short branch with simple leaves and a pinnately compound leaf.

For those who want to know a woody plant at any time of the year I can not emphasize too strongly the importance of twig characters as diagnostic features. The winter buds, indeed, are present not only in the winter, *but all the year through*, with the exception of a short period in May or June when they are unfolding and forming new stems and leaves ("shoots") and perhaps flowers. Then, in May or early June, new buds begin to form in the axils of the new leaves and perhaps at the tip of the stem. These are the winter buds, and in this book will be called simply "buds" (cf. pp. 3–6 on woody plant characters).

*Now in its fourth printing, and called "Guide to trees and shrubs." See no. 20 of references, p. 252.

Other means of identification (besides the flowers) are the leaves, twigs or branchlets, bark, general habit, and any other characters such as remnants of fruit or fruit stalks. It seems hardly necessary to say that in winter the best place to find the leaves is on the ground immediately underneath the tree or shrub. It is true that, on the ground, the leaves of some kinds decay quickly, but others, such as those of the oaks and most maples usually last at least into the next spring. Thus, it should not be necessary to wait for flowers to appear before we can name the species, or at least the genus, of a woody plant, unless perhaps we are dealing with fancy varieties of the lilac, mock-orange, deutzia, cotoneaster, etc.

And so I have tried to present as far as possible the characters of branchlets and winter buds (as well as those of the leaves) which can be found *at any time of the year*. To identify a hickory, for example, one should first look at the buds, the most constant character of all (Pl. VII) and not just one bud, but many, then the bark, and leaves (or leaf remnants on the ground) and especially the nuts — perhaps worm-eaten but nevertheless still giving evidence of their shape and structure — and also the husks of the fruit lying about. With all the evidence at hand we can then make our decision. This is sometimes difficult because, particularly in the case of the hickories, we find that they do not always follow the rules laid down for them in the books. Variation from the prescribed forms is occasional; sometimes this is just natural variation and sometimes, doubtless, due to hybridizing with other species or forms. In case of doubt all one can say is that the plant in question is *near* such and such a species. However, in the great majority of cases the species and varieties run fairly close to the descriptions.

The illustrations, in both plates and text figures, have been drawn by Miss Maud H. Purdy, former Staff Artist of the Brooklyn Botanic Garden, from typical living specimens collected by the author. A few have been drawn from herbarium specimens. Plates I and II, however, are reproductions of photographs by Louis Buhle of the Brooklyn Botanic Garden, and were published in bulletins written by the author for the School Nature League of New York City, now affiliated with the National Audubon Society. It has not been possible to illustrate all species and keep the book within a small compass, but most of the common wild and cultivated kinds are shown.

In the section on "Distinguishing Characters" the following marks and printers' types are used.

Names in **heavy type** are of species native or naturalized (i. e. grow "wild") in the area covered by this guide.

Names starred are of species native or naturalized in North America but found in our area only in cultivation.

Names in ordinary type are of foreign species not naturalized and found only in cultivation in our area. They are included with the native and naturalized species because they are considered relatively important for a general knowledge of our woody plants.

All Latin names of genera and species used elsewhere are in italics.

⁰before a name indicates a shrub.

⁰⁰before a name indicates a vine.

All names not thus marked are of species classed as trees.

The scientific name of each plant is followed by an abbreviation of the name of the botanist or botanists responsible for naming the plant. The explanation of these abbreviations may be found in any of the standard manuals.

As an aid to the pronunciation of the scientific names the syllable to be accented is marked; two accents are used: the grave(`) to indicate the long English sound of the vowel, the acute (′) for the short sound. The Latin names are usually pronounced according to the English method, e. g., *Pìnus* = Pỳ-nus; *ovàta* = oh-vày-ta, etc. I have adopted the recommendation of "Standardized Plant Names" to begin all specific names with a small letter and to use only one "i" in Latin genitives.

The following abbreviations are used for words constantly recurring in the descriptions, in addition to the familiar ones used for feet, inches, directions of the compass, and the various States and countries.

Amer.—American.

br.—branch.

brt.—branchlet.

cult.—cultivated or cultivation.

Eu.—Europe (-an).

fl.—flower or flowering.

fr.—fruit or fruiting.

lat.—lateral.

lf.—leaf.

lft.—leaflet.

nat.—natural.

natzd.—naturalized.

P & G—*Plants and Gardens.*

sp. and spp.—species.

SPN—in parentheses following name, means that this is the name recommended by Standardized Plant Names' Committee.

var.—variety.

yr.—year.

♀—female. ♂—male.

It is a pleasure to have this opportunity of acknowledging the advice, encouragement, and assistance of the many persons who have been interested in this undertaking: especially are my thanks due to Dr. George S. Avery Jr., Director, Hester M. Rusk, Instructor and Asst. Editor, Louis Buhle, photographer, and William E. Jordan, Librarian, all of the Brooklyn Botanic Garden; to Rutherford Platt of New York City, Prof. Joseph Ewan, Tulane University and Prof. Fay Hyland, University of Maine, Mrs. Gladys Gordon Fry, Ornithologist, and Prof. A. W. Evans and Dr. John R. Reeder of Yale University.

WALLINGFORD, CONNECTICUT   A. H. G.
June, 1951

PREFACE TO REVISED EDITION

The enthusiastic welcome with which this book has been received not only in the Northeastern United States but also throughout the country, in Canada, and in many foreign countries has been most encouraging. In response to many requests from horticulturists and from schools of forestry and horticulture, I have prepared a "Winter Key" (19), first published separately but now included with this edition. I have also changed the nomenclature, where necessary, to conform with that of the new "Check List of Native and Naturalized Trees of the U. S." (37), published since the first edition appeared. In the capitalization and compounding of common names I have adopted the plan of hyphenating the parts and beginning the second part with a small letter when the plant in question is not a genuine representative, as e. g., Red-cedar instead of Red-Cedar, Redcedar, or Red Cedar, and Osage-orange instead of Osage-Orange, Osageorange, or Osage Orange.

A few plants have been added in the text, thanks to suggestions from many correspondents. Some of the figures have been replaced by new ones for better detail; some new figures have been added. All these illustrations have been drawn, as in the first edition, by one of the world's foremost botanical artists, Maud H. Purdy, and, as before, have been engraved on copper. In the preparation of the "winter key," I am much indebted for material and suggestions to the following persons: Prof. Fay Hyland, University of Maine; Mr. Louis Buhle, Mr. James Franklin, Dr. Donald G. Huttleston, and Mr. William E. Jordan of the Brooklyn Botanic Garden; Prof. W. H. Camp of the University of Connecticut; Prof. Robert B. Clark of Rutgers University; Dr. Lily M. Perry and Mr. A. G. Johnson of the Arnold Arboretum; Dr. and Mrs. John R. Reeder of Yale University; Miss Hester M. Rusk of Metuchen, N. J.; and Mr. Arthur Waldron of Brooklin, Maine.

Finally, it is a pleasure to have the opportunity here to express my gratitude to the many individuals, not only in this area but throughout the world, who have shown interest in the book and have contributed valuable suggestions toward the making of this revised edition.

WALLINGFORD, CONNECTICUT   A. H. G.
September, 1955

CONTENTS

Illustrated Guide to Trees and Shrubs

INTRODUCTION

Nomenclature. For plant names I have followed Gray's Manual, eighth edition, for the most part, but have used also Rehder's Manual, and Standardized Plant Names, second edition, 1942. The last was prepared by a special committee in 1923 for the purpose of unifying or standardizing the names of the higher plants important in horticulture. This is certainly a laudable object, for the many names for some of our plants have led to much confusion. But this is a big country: even the area covered by this book is about half again as large as that of the British Isles. And so there will be of necessity many common names, or at least more than one, for our common plants, and on this account I have often cited more than one. What we should strive for, however, is to have a uniform single scientific or botanical name. This should serve to identify any given plant with certainty.* I have always given the preferred common name first.

Trees, shrubs, and vines. What dimensions must a woody plant have to entitle it to be classed as a tree? Formerly Sudworth included woody plants having one well-defined stem and a more or less definitely formed crown and attaining a height of at least 8 feet and a diameter of not less than 2 inches. Harlow and Harrar (26, p. 1) have extended the height to 20 feet. Blakeslee and Jarvis (5, p. 434) say 15 feet. I should be inclined to accept the last figure, but the whole matter is of course entirely arbitrary. Shrubs, i. e. woody plants with many more or less erect stems, such as the Lilac, often reach a height of much more than 8 feet, but can not therefore be classed in the category of trees. Japanese potted trees (Brooklyn Botanic Garden Guide no. 6, 1931) have been dwarfed by special methods and are only a few feet in height, but are nevertheless properly called trees. (See also P. & G. 5: no. 3. 1949. and 6: 68–79. 1950.) It would seem that for admission to the class of trees, the habit or form of the plant is of as much importance as the ultimate height.

Vines are either herbaceous or woody, and in the latter case may also be called climbing shrubs. They twine around or fasten themselves to other plants or suitable supports or they trail along the ground. Such are the Grape, Virginia Creeper, Clematis, Poison-ivy etc. They are known technically as lianas and are much more abundant in tropical forests.

---

* For a further discussion of this subject see Moldenke, H. N. American Wild Flowers. (39) pp. XX ff. 1949.

Classification. For the order of families I have used the "natural" system of Engler which is familiar to all botanists. In general, this system depends upon the evolutionary rank of each family—the oldest first and the most recently evolved families last. Thus the Gymnosperms, an ancient group, precede the Angiosperms, which are more modern; and in the Angiosperms the families follow each other in a sequence depending upon their supposed relative ages, with the Compositae, the most modern family of all, at the end.

For the uninitiated, a few words may be in order here about the terms genus, species, etc. The genus is the larger group, as for example; Birch, *Betula*. The species are the *kinds* of birch as *Betula lenta*, Sweet Birch, *B. papyrifera*, Paper B., etc. Species may be further divided into varieties. Nearly related genera, as *Betula*, the Birch, *Carpinus*, the Hornbeam, and *Ostrya*, the Hop-hornbeam, all are classed together in the same family, which in this case is the Birch Family, *Betulaceae*. Nearly related families are grouped together to form orders: as, for example, in the case of the Beech Family, *Fagaceae*, which, with the related *Betulaceae*, belongs to the order *Fagales*. In this book I have omitted the orders, since they are not essential to our purpose. Orders again are grouped into classes, and so on to the main divisions of the plant kingdom from the most ancient to the most modern, as follows:

PLANT KINGDOM*

Division 1. Thallophytes. Bacteria, Algae, Fungi, Lichens.

Division 2. Bryophytes. Mosses and Hepatics.

Division 3. Pteridophytes. Including Clubmosses, Horsetails and True Ferns.

Division 4. Spermatophytes. Seed Plants.

    Subdivision 1. Gymnosperms. (*Gymnos*—naked and *spermon*—seed). Woody plants in which the seeds are *borne naked* on the surface of cone scales. This group includes the Cycads, mostly now tropical or subtropical plants.

    Subdivision 2. Angiosperms. (*Angion*—vessel and *spermon*—seed). Flowering plants. Either woody or herbaceous; seeds enclosed in an ovary (*angion*) as in Orange, Locust, Blueberry etc.

      Class 1. Monocotyledons. Flowering plants with only one cotyledon (seed-leaf) in the embryo. Includes grasses, lilies, irises, orchids, palms, etc.

* A recent classification, which is being rather widely accepted, unites the Pteridophytes and the Spermatophytes into a single group, the Tracheophytes, or vascular plants, as distinct from the lower groups. See Sinnott (57).

Class 2. Dicotyledons. Flowering plants with 2 cotyledons in the embryo. Highest group of flowering plants, some herbaceous and some woody. Most of the plants in this book belong to this group and to the Gymnosperms.

Although some of the earliest botanists centuries ago divided all plants into two classes, namely, herbs and woody plants, it is obvious that the woody plants are not a "natural" group from a systematic viewpoint. In some families, as in the Maple Family, all members are woody; in others, as in the Heath Family, some are woody, some herbaceous. In general, however, the *large* plants of our time are woody. If we make the acquaintance of these, the herbaceous members may soon fall into their proper places.

Woody Plant Characters. During the growing season, woody plants can of course be most readily identified by their leaves. Also important are the characters of the branchlets with their buds, and the bark of the older branches and of the trunk. Flowers and fruits are also very helpful, and sometimes practically necessary; but their absence at certain times of the year and on young specimens makes them less desirable as key characters.

The features of a winter twig are less familiar to most people than those of a leafy stem. For this reason a brief description of a typical winter twig, such as that of the Bigtooth Aspen, is given here. At the tip of the stem is a large bud, the *terminal bud*, A, Fig 1. Along the sides of the stem are other buds, *lateral buds*, cf. *B*, usually smaller, each one situated above a *leaf scar*, *C*. A leaf scar is the mark left where a leaf of the previous summer fell from the stem. The location of buds above the leaf scars shows that these buds were formed in the axils of the leaves; and they are, therefore, called *axillary buds*. Some plants do not have a true terminal bud: in such cases the leaf scar is of course immediately below the "false" terminal bud; and on the other side of the twig we find a short stub or scar indicating the dead end of the season's growth. (Cf. Fig. 103, p. 213.) The leaf scar is marked with several small dots; these are the *vascular bundle scars*, cf. D, marking the places where the conducting strands ("veins") extended from the stem into the leaf. On this stem there is a pair of small scars, one on each side of the leaf scar; these are *stipule scars*, cf. E, showing where these append-

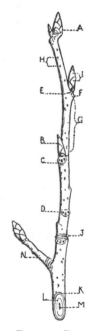

Fig. 1. Brt. of Bigtooth Aspen. For explanation of lettering see text.

ages were attached. (Many plants do not have stipules.) The place on the stem where a leaf is borne (in some plants several leaves) is called a *node*, F, and the part of the stem between two adjacent nodes is the *internode*, G. Scattered over the surface of the stem are small dot-like markings, slightly raised. These are the *lenticels*, *H*, regions of loosely fitting cells with air spaces among them, in the otherwise impervious corky covering of the stem.

A bud is an undeveloped shoot; it consists of a little stem, bearing tiny leaves, or flowers, or both; in woody plants of this climate it usually has its lowest leaves modified into scales, *bud scales*, *I*, which cover and protect the rest of the bud. The arrangement of the bud scales also differs: in the majority of plants they overlap each other, shingle fashion, and are then called *imbricate;* or they may touch each other only at their edges, two or three of them covering the whole bud in this way, an arrangement called *valvate;* or, rarely, only a single cap-like scale may cover the whole bud.

The young stem and leaves within the bud lie dormant during the winter. In the spring the stem begins to lengthen and the leaves to expand, and the scales drop off. By this unfolding of the terminal bud, the stem that bears it becomes longer. The internodes between the expanding foliage leaves usually lengthen considerably, but those between the bud scales lengthen very little or not at all. Hence, when the bud scales fall, they leave a series of scars very close together; the scars themselves are very narrow, and they look like a succession of rings around the stem. A group of such *bud scale scars*, *J*, marks the beginning of each year's growth in length; these are visible on the stem for a number of years, until they are obliterated by its growth in thickness. The age of a twig can be determined by counting the groups of bud scale scars back from the tip. Some of the axillary buds may develop similarly, forming side branchlets, *cf. N*.

But some plants have naked buds; i. e., buds *without* typical scales. In these plants the outermost, leaflike parts of the bud are nevertheless protective and often drop off with the unfolding of the bud.

Most woody plants have *buds of definite growth*, in which all the leaves that are to develop on a given branchlet in one season are laid down in the bud the previous summer. When these leaves have expanded, the stem stops growing in length, and usually forms a terminal bud,* unless the branchlet ends in a flower or flower cluster; a bud is also formed in the axil of each of the leaves. But some plants have *buds of indefinite*

---

* See p. 9, middle, paragraph numbered "1," on "false" terminal buds.

growth, in which there are a few fully formed leaves, and many others just beginning. When such a bud unfolds, the fully formed leaves expand and some of the partly formed leaves finish their development, and buds are formed in their axils. Such a stem continues to grow in length and to put out new leaves until cold weather stops it in the fall; it forms no terminal bud, and its growth in length the next year is taken up usually by the uppermost well-formed axillary bud.

The stem interior is divided roughly into three concentric regions; *bark*, *K*, *wood*, *L*, and *pith*, *M*. The age of any part of the stem may be determined by counting the annual rings in the wood, which are clearly visible in cross section. The age thus determined would, of course, agree with the age computed by counting groups of bud scale scars.

The *phyllotaxy*, or arrangement of leaves on the stem (and consequently the arrangement of axillary buds), is in a general way constant for each species, although there may be some variation on different parts of the same plant, and occasionally some distortion due to twisting of the stem during growth. In identifying a plant by its leaves or buds, it is of the utmost importance first of all to observe its phyllotaxy. The leaves may be arranged in whorls (circles) of three or more at a node (Pl. XLI, 2 and 3, p. 226), or they may be opposite each other on the stem (2 at a node, see Pl. XXVIII, p. 171). In these cases we must consult the key to opposite or whorled leaves or buds. More often, however, leaves, or rather their points of insertion on the stem, are in a spiral arrangement, that is, an imaginary line connecting these points of insertion will be a spiral (Fig. 1a). Then the phyllotaxy is expressed by the distance around the cylindrical stem from one leaf to the next. Thus, if the next leaf of the spiral is halfway around the stem the phyllotaxy is called $\frac{1}{2}$, and the third leaf will be directly above the first, as in Fig. 34, p. 112. More commonly the distance around the stem from one leaf to the next is equal to 2/5 of the stem circumference, and we must pass through five leaves and twice around the stem before arriving at the leaf directly above the one at the starting point. This is 2/5 phyllotaxy (e. g., Pl. XXIV, 1, p. 162 and Fig. 71, p. 155). It will be seen that the denominator of this fraction shows the number of leaves passed before we arrive at the one directly above the starting point, and the numerator the number of turns around the stem. In the bayberry the

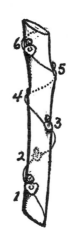

FIG. 1a. Diagram, with 1f. scars connected by an imaginary spiral, showing phyllotaxy.

phyllotaxy is ⅜, and successively higher fractions occur in pine cones, sunflower fruit heads, etc.   Thus we can always ascertain what the next higher phyllotaxy would be by adding the numerators and denominators of the two preceding phyllotaxies; $1/2$, $1/3$, $2/5$, $3/8$, $5/13$, etc.   These relations are obviously the result of geometrical space requirements of the dividing cells at the growing point of the stem where the leaves originate.

HOW TO USE THE KEYS

Before trying to identify a plant we should make sure that the plant in question is neither Poison-ivy nor Poison Sumac, for most people can not handle these plants without being poisoned. The rash resulting from poisoning usually does not appear for about a day. These poisonous plants will be found illustrated on Plate XXV, 1 and 2, and described on pp. 163 and 165. Drawings of the branchlets in the winter condition are reproduced herewith. (Fig. 2.)

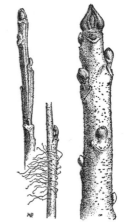

FIG. 2. Brts. of Poison-ivy (left) and Poison Sumac. For description see pp. 163 and 165.

Before attempting to use the keys it is best to have a hand lens of about 10 × power available to determine fine points such as pubescence, veining etc. However, the keys have been so arranged as to make the lens usually unnecessary.

By means of the "keys" the name, or at least the genus, of any woody plant growing in the area covered by this guide can be determined. There are a few exceptions; extremely rare plants or plants which occur only on the borders of our area have usually been omitted from the keys, but for the most part will be found briefly described in the text.

First of all, we must be certain that the plant to be identified is *woody*, i. e., either tree, shrub, or woody vine; for this book deals only with woody plants. All other plants, such as violet, dandelion, daisy, etc., are herbs (herbaceous) and contain no wood (using the term in the popular sense) in their tissues.

Next, we must find out to which of the following general groups of woody plants our specimen belongs.

   I. Gymnosperms. Seeds naked. Lvs.* needle-shaped, scale-like, or linear. Pine and Yew Families; including also these Angiosperms: Heath, *Erica;* Heather, *Calluna;* Dwarf-mistletoe, *Arceuthobium;* and Tamarisk, *Tamarix.* p. 13.

   II. Angiosperms. Seeds in an ovary. Broad-lvd. woody plants.

       A. With opposite or whorled evergreen lvs. p. 14.

       B. With opposite or whorled deciduous lvs. p. 15.

       C. With alternate evergreen lvs. p. 18.

       D. With alternate deciduous lvs. p. 20.

* The word "leaves" (lvs.), used frequently in the keys, should have a word of explanation. Strictly, a typical leaf is composed of blade, petiole, and stipules. But it is understood that in the keys "leaves" means leaf *blades*.

The "summer key," to be used at the time when woody plants are in leaf, or during the season of active growth of plants, begins on p. 13. The "winter key" is to be used when woody plants are leafless and growth is dormant. This begins on p. 30. Both these keys are of similar construction, being dichotomous, or with only two choices, except in a few instances where the choices (three or more) are mutually exclusive.

Let us assume that the plant to be identified has been collected during the leafy season of woody plants, or, generally speaking, during the period from about May 1 to November 1. Therefore we turn to the "summer key." Let us assume next that, after investigation, we find that the plant belongs to "D" of II, Angiosperms, that is, of broad-leaved woody plants. Turning to "D," p. 20 of the summer key, we find that No. 1, the figure at the left, reads "Lvs. compound." Since in our plant the leaves are simple we turn to the second "1" near the top of p. 22. (In case the terms "compound" and "simple" are not clear they will be found defined, as are all unfamiliar terms, in the Glossary, p. 255. This should be used, especially by beginners, unsparingly.) Under the second "1" we find "36. Lvs. mostly lobed," and finding that our leaves are *not* lobed we turn, as directed, to the second "36" near the middle of p. 23. "Lvs. mostly not lobed." Under this "36" we find three "60's," i.e., "Lvs. parallel veined," "Lvs. palmately veined," and "Lvs. pinnately veined." We determine that our leaf belongs under the second "60." Now we are given another pair of choices as to whether the plant in question is a climbing vine or is not climbing. Having decided on the latter, for we know it is a large tree, we find below two "64's," "sap milky" and "sap not milky." To determine this point we cut across a twig or a leaf petiole. If the sap is milky, it will appear as white drops in the cut section. Finding that the sap is not milky we next take the first "66," which states that the "leaves are somewhat fleshy, the lower ones opposite and that it is a shrub of salt marshes." None of these things applying to our plant, we take the second "66" and are led to decide as to whether the leaves are toothed or not ("67's"). Seeing that the leaves are toothed and, moreover, have tufts of hairs in the axils of the veins beneath, we are led to *Tilia*, the Linden or Basswood. Next we confirm our decision by referring, as directed, to Pl. XXIX, 1 and 2, p. 178 and to the description, pp. 179, 180. Where there are many species of a genus there is usually a key to these, which should now be used to determine the species. In some cases, however, only a brief review of the species is given and from this we can probably determine our species.

The winter key, for plants from November to May, is constructed and used in a similar way. If, for example, we should have a winter twig of *Tilia*, we would be led, assuming that we had the American

Linden or Basswood, after an inspection of the classes of "D," pp. **34, 35,** to No. 113 "Brts. with distinct coloring etc.," for at least the American Linden has a bright green or red brt.   Under 113 we would find the cor: rect answer in the third 114, p. 45, since the buds are lopsided.   Or, if we should have some other species of *Tilia without* pronounced color we should select the second 120, p. 47, "True terminal bud absent." Under this number we should finally arrive at the second 146, p. 48: "Buds with not more than 3 scales exposed" and would again be brought, because of the *size* of the bud, to *Tilia*, the first 151 near the middle of p. 48.

"If at first you do not succeed, try, try again."   You may have made a wrong choice somewhere and so should start over again.   In every case at least a tentative decision between two groups must be made; but the possibility of mistake in judgment must always be kept in mind; and in case of doubt both groups must be tried.

Short Cuts.   Some plants have one or more particularly striking characters such as chambered pith, large terminal buds etc. by which they can be instantly recognized.   These I have placed in special groups at the end of the part on "Distinguishing Characters," p. 243.   If our plant is not in these special groups there is no other recourse but to use the longer general key.

In anticipation of some of the difficulties in using the keys, the following suggestions are given:

1. To determine the presence or absence of a true terminal bud one must examine the tip of the branchlet carefully with the lens.   *Generally speaking*, if the topmost bud is situated just above a leaf scar, it is axillary, and a true terminal bud is lacking.   However, there must be something to represent the true stem tip (at another side of the bud, away from the leaf scar); either a small stub, or another scar—*different from the leaf scars*.   Ordinarily, above each leaf scar there is a bud, large or small.   But where several leaves were close together near the tip of the stem, there may be two or more leaf scars (near the terminal bud) without buds above them, as often in the Wild Cherry (*Prunus serotina*).

In trees like the Hickory, Tulip-Tree, and most conifers, the fact that the end bud is a true terminal bud is obvious: in the oaks, too, this is the case, although in the oaks as well as in the conifers it is sometimes hard to pick out from the cluster of buds crowded at the end of the branch, which one is the true terminal bud, but it is usually the largest one.

2. A young tree may look like a shrub; and therefore what appears to be a shrub may have to be sought under "trees."

3. A climber or vine is not always easy to distinguish in the very young stage if it lacks tendrils or has not yet started to climb; but the

branchlets are usually long and very slender, and may early show a tendency to coil.

4. Whether the leaves are opposite or whorled, or alternate is sometimes puzzling if the internodes are very short, as in the Mountain-Laurel (*Kalmia latifolia*) and in some of the conifers. In case of uncertainty, it should be assumed that the leaves are alternate, as pairs and whorls of leaves usually show fairly distinctly, especially if the stem is viewed from the tip rather than from the side. The leaves of any pair or whorl are usually of approximately the same size; hence if leaves appear to be at the same level but are markedly different in size, the probability is that the small one is actually younger and nearer the tip of the stem. A notable exception occurs in Catalpa, in which there are usually two large leaves and one small one at a node. Leaf scars on older parts of the twig often show their arrangement more clearly than the leaves of the current season.

5. *A compound leaf can be distinguished from a branchlet with several simple leaves by the presence of a bud in its axil, and by the lack of buds in the axils of its leaflets; it may be terminated by a leaflet, but never by a bud. A branchlet has no bud in its axil, but has a bud in the axil of each of its leaves; it may be terminated by a bud, but never by a leaf. Cf. figs. 34, p. 112, and 107, p. 220. A deeply cut leaf is not considered compound unless it is divided into separate parts all the way to the rachis or midrib.*

6. Plants classed as having "lvs. symmetrical at base" often have some unsymmetrical leaves; and even on plants characterized by having "lvs. very unsymmetrical at base," some of the leaves may be nearly symmetrical. One must look at *many leaves*, therefore, and try to judge the plant fairly, by the majority.

7. The difference between shallow lobes in a leaf and large, coarse teeth may not always be clear, in such plants as the American Holly (*Ilex opaca*) and Swamp White Oak (*Quercus bicolor*). One should be ready to change his opinion if he has trouble in naming his plant.

8. By midsummer the buds for the following season are well formed in the axils of the leaves and at the tips of the branchlets, and are typical in appearance. Early in the summer, however, when the buds are not fully developed, they may not look typical or may be hardly visible; at this time, therefore, one may have to look for dormant buds on twigs of the previous year.

9. Evergreens may be recognized in the summer by the presence of leaves produced in more than one season. Leaves of the current season are usually much lighter in color. The current year's growth of the stem, too, can usually be recognized by a difference in color, and by the position (at its base) of the last group of bud scale scars. If leaves are still present on older parts of the stem, the plant is evergreen; but very short branch-

lets of the current year (bearing new leaves) may occur along the sides of older parts of the stem; and the leaves on these may at first appear to be borne directly on the old wood. As a general rule, the leaves of evergreens are of firmer texture than those of decidous plants; at least one can be sure that very thin-leaved plants are not evergreen.

10. As to thorns, prickles, spines, and bristles I have reserved the name "thorn" for those outgrowths from the stem that are in reality organs, i. e. branches, leaves or stipules, that have been modified in the course of evolution into a simple or branched thorn, as in Hawthorn, modified branch; Black Locust, and Toothache-tree, modified stipules; and Barberry, modified leaves. All other sharp pointed outgrowths, as in Blackberry, Roses etc. I have called prickles. The latter all come under the head of "emergences."

11. Most of the conifers have two kinds of leaves: (1) the scale leaves, which are often without a green color, and (2) the green foliage leaves. These are known as the primary and secondary leaves, respectively. The primary leaves are usually small and scale-like and function as bud scales. But in the Pine, in the young seedling stage, the primary leaves are the only leaf form (above the cotyledons) and there resemble and function as foliage leaves. The secondary or needle-like leaves arise from buds in the axils of the upper primary leaves during the first or second years of the seedling. Thereafter the green primary leaves gradually, in subsequent years, assume the form of brownish scales. At length these primary leaves function only as bud scales in the older tree; and, at the time of the unfolding of the winter buds, may be seen at the base of each fascicle of needles. Soon thereafter they may fall off or are at least inconspicuous.

In the Cypress tribe (*Cupressineae*) there is only one kind of leaf, namely the primary, which is scale-like but green and sometimes of two forms in the mature tree, as in Thuja where the leaves on the edges of the shoot are keeled, those on upper and lower surfaces flat.

12. The Pine, Larch, Golden-larch, Cedar, and Bald-cypress have, as a result of the situation described above, long growths and short growths (German *langtriebe* and *kurztriebe*). The short growths are also known as short shoots or spur-like growths. For the fascicle of needles arising in the Pine in the axil of the scale leaf is in reality a shoot whose terminal bud habitually dies or, rather, aborts. But in Larch and Golden-larch and Cedar this terminal bud in the short growth does not abort but furnishes a cluster of leaves each year. In some cases indeed it changes into a long growth, and examples of this are common in any Larch tree.

The long and short growths are only devices for specialization.   The short growths furnish the foliage leaves for photosynthesis while the long growths advance the length of the tree or branch and thus enable it and the tree as a whole to increase its operations.   Although often not as well defined as they occur in the conifers, long and short growths may be found in almost any tree when it becomes mature.   Especially prominent are they in Sweet or Yellow Birches where the short growths furnish the clusters of leaves along the branches.   They are also prominent in "fruit" trees such as the Apple or Pear, where they serve to bear flowers and fruit.

SUMMER KEY

To be used in the leafy season.

(Since the leaves of most gymnosperms are evergreen, this part of the Summer Key, pp. 13, 14 can be used also for gymnosperms of the Winter Key, p. 30.)

I. Gymnosperms

Seeds naked, borne usually in cones (Fig. 15). Pine and Yew Families, including also these Angiosperms: Heath, *Erica;* Heather, *Calluna;* Dwarf-mistletoe, *Arceuthobium;* and Tamarisk, *Tamarix.* Lvs. needle-shaped, awl-shaped, scale-like, or linear, mostly evergreen. Resin ducts present in wood of *Pinus, Pseudotsuga, Larix,* and *Picea.* Resin ducts or glands in lvs. of all spp. except *Taxus.*

Numbers at extreme right refer to pages

1. Lvs. either in pairs or in 3's around the stem, or (at least some of them) in clusters, or fascicles.
  2. Lvs. in alternating pairs (decussate) or in 3's around the stem (Figs. 19–23, pp. 69–72).
    3. Lvs. in alternating pairs, at least some of them scale-like.
      4. Scale-like lvs. forming 4-sided brts. (sometimes awl-shaped lvs. on same plant) (Fig. 22, p. 71)............................*Juniperus* 71
      4. Lvs. all scale-like.
        5. Plants parasitic on conifers, dwarf, not more than 1 in. high; lvs. small brown scales......................................*Arceuthobium* 117
        5. Not parasitic, lvs. forming flat or cord-like brts. sometimes lf. pairs separated in *Calluna.*
          6. Conifers, i. e., cone-bearing trees or shrubs.
            7. Cones globular, cone scales thick; lvs. whitened beneath in exotic spp. (Fig. 21, 1, c, p. 70).....................*Chamaecyparis* 69
            7. Cones conical, cone scale thin; lvs. usually not whitened beneath (Fig. 19, c, p. 69)...............................*Thuja* 68
          6. Angiosperms; plants with small, showy fls.; lvs. appressed to stem, with 2 prongs at base...............................*Calluna* 207
    3. Lvs. in alternating pairs or in 3's, at least some of them awl-shaped or linear.
      8. Conifers; lvs. stiff and sharp either in pairs (with scale-like lvs. forming 4-sided brts. on same plant) or in 3's (Fig. 22, p. 71).........*Juniperus* 71
      8. Angiosperms; plants with small showy fls.; lvs. stiffly outstanding, linear (Fig. 100, p. 209)...............................*Erica* 207
  2. Lvs., at least some of them, in special clusters or fascicles of 2 or more (Figs. 12, 13, 14, 16, pp. 60, 61, 64, 66)
    9. Lvs. 2–5 in each fascicle, usually sheathed at base (Fig. 14, p. 64 and Pl.II)
                                                    Pinus 62
    9. Lvs. many in each whorl-like cluster.
      10. Lvs. 3–5 in. long, in apparent whorls around the stem, each lf. in the axil of a little scale (Fig. 16, p. 66).....................*Sciadopitys* 67
      10. Lvs. less than 3 in. long, in apparent whorls on short thick spurs, some lvs. also arranged spirally along long growths of stem.

11. Lvs. evergreen, stiff, sharp (Fig. 12, p. 60)........................*Cedrus* 61
11. Lvs. deciduous.
    12. Lvs. needle-like, soft and blunt, fr. a cone.
        13. Lvs. short, ½–1½ in. long; cone scales persistent................*Larix* 60
        13. Lvs. longer, ¾–2 in. long; cone scales deciduous...........*Pseudolarix* 59
    12. Lvs. fan-shaped, often lobed, seed drupe-like (Fig. 3, p. 50).........*Ginkgo* 50
1. Lvs., at least some of them, spirally arranged (Fig. 5, p. 54).
    14. Lvs. needle- or awl-shaped.
        15. Lvs. mounted on little stalks, 3- or 4-sided.
            16. Lvs. 4-sided, all spirally arranged, set close together on brts. (Figs. 5, d,
                9, 10, pp. 54, 56, 57, and Pl. I)..............................*Picea* 56
            16. Lvs. 3-sided, sparsely scattered on long growths of stem, apparently
                whorled on short spurs (Fig. 12, p. 60).......................*Cedrus* 61
        15. Lvs. not mounted on little stalks.
            17. Lvs. strongly decurrent on stem, conifer (Fig. 18, p. 68)......*Cryptomeria* 68
            17. Lvs. often sheathing stem but not decurrent; angiosperm with tiny
                showy fls. in dense, slender spikes; leafy brts. often deciduous...*Tamarix* 182
    14. Lvs. linear, flattened.
        18. Lvs. abruptly narrowed on short stalks, all spirally arranged and mounted
            on small shoulders which are decurrent on stem.
            19. Lvs. sharp-pointed at tip (Fig. 4, p. 52).......................*Taxus* 51
            19. Lvs. blunt at tip (Figs. 5, c, 7, and 8, pp. 54 and 56)...........*Tsuga* 55
        18. Lvs. not abruptly narrowed into distinct stalks.
            20. Lvs. deciduous, soft.
                21. Lvs. very short, up to ½ in., apparently 2-ranked on the short growths,
                    making these resemble compound lvs. (Fig. 17, p. 67).....*Taxodium* 67
                21. Lvs. longer, ½–1½ in., spirally arranged on long growths, in apparent
                    whorls on short growths.
                    22. Lvs. short, ½–1¼ in., short growths short, cone scales persistent
                        (Fig. 11, a, p. 60).......................................*Larix* 60
                    22. Lvs. a little longer, ¾–2 in.; short growths longer (Fig. 11, b, p. 60)
                            Pseudolarix 59
            20. Lvs. evergreen, firm, all spirally arranged, leaving circular or elliptical
                lf. scars.
                23. Lf. scars circular, flat against the brt. (Fig. 5, a, p. 54).........*Abies* 53
                23. Lf. scars elliptical, raised from the brt. at the lower edge (Figs. 5, b,
                and 6, pp. 54, 55)................................*Pseudotsuga* 54

II. Angiosperms. Seeds in an ovary.

A. Broad-lvd. Woody Plants with Opposite or Whorled Evergreen Lvs.

Lvs. all simple.

1. Climbing vines.
    2. Lvs. serrulate, in some vars. variegated.....*Euonymus fortunei.* var. *radicans* 170
    2. Lvs. entire; 1 or 2 pairs below fls. connate (Pl. XLV, 3)..*Lonicera sempervirens** 240

---

* Lvs. evergreen only toward the south of our area.

1. Not climbing.
  3. Lvs. linear or scale-like; low heath-like shrubs.
    4. Lvs. scale-like, opposite but in 4 ranks lengthwise on stem........*Calluna* 207
    4. Lvs. linear, in whorls of 3 or 4 (Fig. 100, p. 209)..................*Erica* 207
  3. Lvs. not as above.
    5. Prostrate or creeping.
      6. Lvs. ovate to oblong ($\frac{1}{3}$–1 in. long); fls. large (about 1 in.), blue; juice
        milky in young shoots.....................................*Vinca* 224
      6. Lvs. rounded, small ($\frac{1}{4}$–1 in. long); subshrubs; fls. smaller, not blue.
        7. Lvs. toothed; fr. dry (Pl. XLIV, 3)...*Linnaea borealis* var. *americana* 238
        7. Lvs. entire, or undulate; fr. a double berry.................*Mitchella* 230
    5. Not prostrate or creeping.
      8. Half parasitic on other woody plants; southern native; lvs. thickish,
        entire (Fig. 35, p. 116)..............................*Phoradendron* 117
      8. Not parasitic.
        9. Lvs. small ($\frac{1}{2}$–1 in. long), cult.
          10. Lvs. somewhat leathery, rounded or emarginate at apex; shrub
            much used for hedges (Fig. 75, p. 160)...................*Buxus* 160
          10. Lvs. thinner and softer, pointed at apex, hairy near base of midrib,
            half evergreen (Pl. XLI, 4)...................*Abelia grandiflora* 238
        9. Lvs. averaging larger (1–2$\frac{1}{2}$ in. long), native.
          11. Lvs. often in whorls of 3; end bud abortive; low native shrub of
            woods (Pl. XXXV, 3).....................*Kalmia angustifolia* 200
          11. Lvs. opposite only; end bud distinct; much cult. for hedges; half
            evergreen (Pl. XL, 3)..........*Ligustrum vulgare* and *ovalifolium* 221

B. Broad-lvd. Woody Plants with Opposite or Whorled Deciduous Lvs.

1. **Lvs. compound.***
  2. Lfts. 3 (3-foliolate).
    3. Woody vines; climbing by means of twining petioles............*Clematis* 118
    3. Shrubs; not climbing.
      4. Bark greenish and striped; fr. bladdery; lfts. 3, rarely 5; native (Pl.
        XXVII, 1)...*Staphylea* 170
      4. Bark not as above; fr. a 2-celled, dry capsule; lvs. usually simple, some-
        times 3-foliolate (Pl. XL, 1, b)............................*Forsythia* 221
    3. Trees; lfts. 3–5 (Pl. XXVIII, 3)........................*Acer negundo* 174
  2. Lfts. more than 3.
    5. Lfts. palmately arranged.   (Cf. fig. 80, p. 175.)
      6. Aromatic shrubs, to 9 ft.; buds small, yellow-brown, woolly, rounded,
        often superposed; lfts. 5–7, short stalked, 2–4 in. long...........*Vitex* 225
      6. Large trees, rarely shrubs; buds large, smooth or resinous, pointed, not
        superposed; lfts. 5–7, not stalked (Fig. 80, p. 175).............*Aesculus* 176
    5. Lfts. pinnately arranged.
      7. Vine; climbing by aerial roots (Fig. 111, p. 229).......*Campsis radicans* 227
      7. Shrubs; pith wide and soft; fr. black or red berries (Pl. XLIV, 1 and 2)
                                           Sambucus 231
      7. Trees.
        8. Lf. scars horseshoe-shaped; fr. a cluster of black drupes; lfts. 5–13 (Pl.
          XXIII, 1)...*Phellodendron* 160

---

* The most important divisions of the longer keys are printed in bold face type.

8. Lf. scars and fr. not as above.
    9. Fr. a double samara; lfts. 3–5 (Pl. XXVIII, 3)............*Acer negundo* 174
    9. Fr. a single samara; lfts. 5–13 (only 1 in *F. anomala*) (Pl. XXXIX). *Fraxinus* 219

1. **Lvs. simple.**
  10. Lvs. mostly lobed.
    11. Brts. hollow or with chambered or diaphragmed pith; lvs. often not lobed.
      12. Lvs. large, 5–10 in. long, very shallowly 3–5-lobed, or undivided; tree
        (Pl. XLII)...*Paulownia* 227
      12. Lvs. smaller, 2½–5 in. long, deeply 2- or 3-parted or undivided; serrate or
        occasionally entire; shrubs (Pl. XL, 1, b and c)............*Forsythia* 221
    11. Brts. with solid pith.
      13. Lvs. irregularly lobed, or undivided.
        14. Lvs. rough above, soft-pubescent beneath, sometimes alternate; pith
          with a thin green diaphragm at each node (Pl. XVII, 3)...*Broussonetia* 114
        14. Lvs. glabrous above, pubescent beneath, usually entire, but sometimes
          with 1 or 2 lateral teeth, and in 3's, never alternate (Pl. XLI, 3)
                                          Catalpa 227
      13. Lvs. palmately 3–11-lobed; all lvs. opposite.
        15. Lvs. 3-lobed; fr. red or black drupes; shrubs.
          16. Fr. small black drupes in clusters (Fig. 112, p. 233)
                              Viburnum acerifolium 232
          16. Fr. red drupes in clusters.........*Viburnum trilobum* and *opulus* 232
                              See key to Viburnum, p. 232
        15. Lvs. 3–11-lobed; fr. a double samara; trees (except *A. spicatum*, some-
          times a shrub) (Pl. XXVIII and Figs. 78, 79, 79a, and 79b)....*Acer* 172
  10. **Lvs. mostly not lobed.**
    17. Sap milky; lvs. often lobed, sometimes alternate (Pl. XVII, 3). *Broussonetia* 114
    17. Sap not milky.
      18. Lvs. averaging large, 3–12 in. long.
        19. Lvs. entire; trees.
          20. Brts. pubescent; lvs. ovate or ovate-oblong, pointed, 3–6 in. long;
            tree or large shrub (Pl. XL, 4, and Fig. 109, p. 223)...*Chionanthus* 223
          20. Brts. glabrous, at least when old.
            21. Lvs. pubescent above and below, regularly opposite, sometime
              shallowly 3–5-lobed (Pl. XLII)...................*Paulownia* 227
            21. Lvs. glabrous or nearly so above, more often in whorls of 3, usually
              not lobed (Pl. XLI, 3)..............................*Catalpa* 227
        19. Lvs. serrate; mostly cult. shrubs.
          22. Brts. more or less 4-sided, at least toward tip, dying back in winter;
            lvs. white-tomentose beneath, long-pointed (4–10 in.).....*Buddleia* 223
          22. Brts. terete; lvs. not as above, 2–8 in. long; bark exfoliating on older
            brts.; lvs. often in whorls of 3.......................*Hydrangea* 128
      18. Lvs. averaging smaller, not more than 4 in. long.
        23. Lvs. entire.
          24. Lvs. with silvery and brown scales beneath, sometimes slightly so
            above; see also *S. argentea*, p. 187.........*Shepherdia canadensis* 186
          24. Lvs. not as above.

25. Fls. peculiar; calyx and corolla similar, of many red-brown segments with odor of strawberries when crushed; wood with odor of camphor; lvs. ovate or narrow-elliptic, soft downy beneath, 2–5 in. long (Fig. 44, p. 124). .*Calycanthus* 124

25. Fls. not as above.

  26. Fr. a drupe.

    27. Lvs. small, 1–2½ in.; bud scales usually imbricated (Pl. XL, 3, and Fig. 108, p. 223). . . . .*Ligustrum* 221

    27. Lvs. *averaging* larger; bud scales nearly or quite valvate (Pls. XXXI and XXXII). . . . .*Cornus* 189

  26. Frs. very small and dry, densely crowded in spherical heads; lvs. sometimes in 3's or 4's; buds sunken in bark; native shrub of swamps or wet soil (Pl. XLI, 2). . . . .*Cephalanthus* 230

  26. Fr. a berry or berry-like drupe; brts. sometimes hollow.

    28. Berry white or red, with 2 seeds; upright shrubs; lvs. sometimes lobed
        Symphoricarpos 238

    28. Berry red or black, with many seeds; berries often in pairs, sometimes fused, or in axillary clusters; upright or twining shrubs (Pl. XLV)
        Lonicera 240

  26. Fr. a capsule.

    29. Capsules glabrous; lvs. heart-shaped or oval, glabrous in our spp. (Pl. XL, 2). . . . .*Syringa* 221

    29. Capsules bristly, small; lvs. remotely or shallowly toothed, ciliate, hairy at least on veins beneath. . . . .*Kolkwitzia* 238

23. Lvs. toothed.

  30. Lvs. small, averaging less than 1 in. long; sepals leaf-like, purplish, persistent; handsome, long-flowering, half-evergreen shrub, with white or pinkish fls.; cult. (Pl. XLI, 4). . . . .*Abelia* 238

  30. Lvs. longer than 1 in.

    31. Lvs. oblanceolate, often alternate, toothed above; buds with a single, cap-like scale (Pl. III, 3). . . . .*Salix purpurea* 78

    31. Lvs. and buds not as above.

      32. Lvs. somewhat fleshy, the upper ones alternate; shrubs of salt marshes (Fig. 116, p. 242). . . . .*Iva* 242

      32. Lvs. not as above.

        33. Lvs. with rounded shallow teeth, round heart-shaped (Fig. 41, p. 122)
          Cercidiphyllum 120

        33. Lvs. with pointed teeth, serrulate, serrate, dentate, or occasionally entire.

          34. Brts. hollow or with chambered pith.

            35. Lvs. sometimes 3-parted or with 3 lfts., occasionally entire; fls. yellow; bark not exfoliating (Pl. XL, 1). . . . .*Forsythia* 221

            35. Lvs. always simple, often rough above; bark usually exfoliating; brts. hollow; fls. white or pinkish (Fig. 47, p. 127). . . . . . .*Deutzia* 127

          34. Brts. not as above; solid.

            36. Brts. some shade of green, often winged; fr. a 4-lobed capsule, sometimes divided to the base; seed with red or pink aril (Fig. 77a, p. 169, and Pl. XXVII, 4). . . . .*Euonymus* 169

            36. Brts. when mature, some shade of brown.

37. Brts. (long-growths) ending in short thorn; fr. a 2–4 seeded drupe; fls. and
frs. in axillary clusters; lvs. sometimes alternate; shrubs or small trees (Fig.
82, p. 177)..*Rhamnus cathartica* 176

37. Brts. not as above, shrubs.

38. Lvs. doubly serrate, with stipules; fr. consisting of 4 dry, shiny drupes (Fig.
64, p. 143).................................*Rhodotypos scandens* 142

38. Lvs. dentate, serrate, serrulate or sometimes entire, without stipules.

39. Winter buds hidden under base of petiole, sometimes free; lvs. dentate
or nearly entire; fls. white with 4 separate petals, fragrant (Figs. 45, 46,
p. 126)...*Philadelphus* 127

39. Winter buds free.

40. Brts. pubescent; a low ridge extending downward on brt. from median
point between lvs.; lvs. elliptic, 2–4 in. long, serrate, pubescent; fls.
rather large, funnel-shaped, pinkish white to crimson (Pl. XLV, 1)
Weigela 238

40. Brts. glabrous or pubescent; no ridges present as above; lvs. serrate,
serrulate, or sometimes entire, lobed in some spp.; fls. small in terminal
clusters; native and cult. spp. (Pl. XLIII, and Figs. 112, 113, 114, pp.
233, 234, 235)..*Viburnum* 231

C. Broad-lvd. woody plants with alternate *evergreen* lvs.

1. Lvs. compound.

2. Lfts. 3, rarely 5; prostate, native, half-evergreen shrub with bristly stems; wet
places..*Rùbus híspidus* 144

2. Lfts. 5–9; upright, cult. shrub; lvs. spiny........................*Mahònia* 119

1. Lvs. simple.

3. Vines.

4. Brts. prickly; lvs. more or less parallel-veined (Fig. 24, p. 73)....*Smilax* spp.   73

4. Brts. without prickles; lvs. net-veined (Fig. 91, p. 189)............*Hédera* 188

3. Trees or shrubs.

5. Lvs. and/or brts. prickly, spiny or thorny.

6. Lvs. spiny; brts. not thorny or spiny (Pl. XXVI, 1)
Ìlex opàca and *aquifòlium* 167

6. Lvs. not spiny; brts. thorny.

7. Fr. small, orange or red pomes, in clusters (Fig. 54, p. 136)
Pyracántha 135

7. Fr. large (1–3 in.), yellow or yellow-green, fragrant (Fig. 61, p. 140)
Chaenomèles 141

5. Lvs. and/or brts. not as above.

8. Lvs. small, not more than ⅔ in. long, needle-like or scale-like.

9. Lvs. needle-like, in whorls of 3 or 4 (Fig. 100, p. 209).........*Erìca* 207

9. Lvs. scale-like, opposite, in 4 rows lengthwise on stem........*Callùna* 207

8. Lvs. larger, but averaging not more than 1 in. long, entire; low spreading
shrubs.

10. Fr. a berry (containing many small seeds).

11. Berries white, aromatic; small, trailing sub-shrub of bogs or mossy
woods; northern, or s. in uplands..*Gaulthèria* (*Chiógenes*) *hispídula* 207

11. Berries some shade of red; small prostrate shrubs; lvs. oval, elliptical
or obovate.

12. Plants of dry, rocky or peaty, acid soil; northern (Pl. XXXVIII, 5, a, b)
Vaccìnium vitis-idàea, var. minus 212

12. Plants of bogs, not limited to northern regions or uplands (Pl. XXXVII, 5)
Vaccìnium oxycóccus and macrocárpon 212

10. Fr. a drupe.

  13. Stone composed of 5–10 wholly or partly fused nutlets; brts. glabrous or nearly so; drupe dull red, dry, and mealy, lvs. obovate, blunt, or rounded at tip, somewhat leathery; native (Pl. XXXVII, 6).......<i>Arctostáphylos</i> 207

  13. Seed single; drupe yellowish brown; young brts. pubescent; lvs. obtuse and mucronulate, softer than in the last; cult.............<i>Dáphne cneòrum</i> 183

10. Fr. a small pome; lvs. orbicular, pointed at ends, $\frac{1}{4}$–$\frac{1}{2}$ in. long; half-evergreen; brts. horizontally branched (Fig. 53, p. 135)........<i>Cotoneáster horizontàlis</i> 135

8. Lvs. averaging from 1–2 in. long.

  14. Taller, upright shrubs; lvs. toothed toward tip (Pl. XXVI, 2, 4)
Îlex glàbra (half-evergreen) and I. crenata 167

  14. Lower-growing shrubs, rarely more than 3 ft. in height.

    15. Lvs. usually opposite or whorled (Pl. XXXIV, 3)....<i>Kálmia angustifòlia</i> 200

    15. Lvs. alternate.

      16. Plants of bogs or pond margins; fls. white, urn-shaped or cylindric, in the axils of small upper lvs. (Pl. XXXIV, 3).....<i>Chamaedáphne calyculàta</i> 204

      16. Plants of drier soil; low or trailing.

        17. Lvs. pubescent; trailing plant (fls. in early spring, fragrant, white to pink (Pl. XXXIV, 4)..........................<i>Epigaèa rèpens</i> 202

        17. Lvs. glabrous; low plants (to 6 in.) fr. a berry, composed in part of the fleshy calyx; whole plant with wintergreen flavor (Pl. XXXV, 4)
Gaulthèria procúmbens 207

8. Lvs. averaging more than 2 in. long.

  18. Lvs. either coarsely or finely toothed.

    19. Lvs. coarsely toothed; low shrub used for ground cover; half-evergreen (Fig. 74, p. 160)............................<i>Pachysándra terminàlis</i> 160

    19. Lvs. more finely toothed.

      20. Lvs. clustered near tip of brt.; early-fl. ornamental shrubs; fls. of blueberry type, white, in long clusters (Pl. XXXV, 1, 2)............<i>Pieris</i> 201

      20. Lvs. not as above—to 6 in. in length, very long-pointed; plant from southern mts.; much used in landscape work (Fig. 99, p. 206)
Leucothoë editòrum 206

18. Lvs. entire.

  21. Brts. aromatic; encircled by stipule scars at nodes (Pl. XVIII, 2)
Magnòlia virginiàna and M. grandiflòra 123

  21. Brts. not as above.

    22. Fl. buds much larger than lf. buds but of same general structure
Evergreen spp. of Rhododéndron 198

    22. Fl. buds in specially formed clusters at tip of brts. (Fig. 96, p. 200)
Kálmia latifòlia 200

D. Broad-lvd. Woody Plants with Alternate, Deciduous Lvs.

(Lvs. only one at each node)

1. **Lvs. compound** (2nd 1, p. 22).
   2. **Stems and lvs. without prickles or thorns** (2nd 2, p. 21).
      3. Lvs., some or all of them, with 3 lfts.
         4. Lfts. all stalked, the terminal one distinctly so.
            5. Lfts. entire, obtuse and mucronate at tip (Fig. 72, p. 155)
            Laburnum anagyroides 153
            5. Lfts. lobed, deeply toothed, merely serrate or sometimes entire, the terminal one with longer stalk.   Vines.
               6. Tendrils present, growing from nodes, flattened at tip into disk-like holdfasts; aerial rootlets occasional; lvs. usually simple but sometimes with 3 lfts.; cult. vine; fr. blue-black, glaucous (Pl. XXX, 1)
               Parthenocissus tricuspidata 179
               6. Tendrils absent; aerial roots abundant; fr. white or whitish; very common native vine.   *Poisonous to touch* (Fig. 2, p. 7) (Pl. XXV, 2)
               Rhus radicans 165
               See also *R. toxicodendron* (southern sp.)
         4. Lfts. not or very slightly stalked.
            7. Lfts. of nearly same size, deeply toothed, pubescent; fls. buds much larger (than lf. buds), containing catkins for next yr.; fls. in March and April; fr. red; rare or local (Fig. 76, p. 163).........*Rhus aromatica* 163
            7. Terminal lft. larger and longer, showing translucent dots when held to light; fr. a winged samara (Pl. XXIII, 3)............*Ptelea trifoliata* 159
      3. Lvs. with more than 3 lfts.
         8. Vines; lvs. palmately compound, of 5 lfts.
            9. Stems with tendrils; lfts. serrate (Pl. XXX, 2)
            Parthenocissus quinquefolia 177
            9. Stems without tendrils; twining; lfts. entire, emarginate (Fig. 38, p. 118)
            Akebia 118
         8. Shrubs or trees; lvs. once pinnate (cf. Pl. XXIV).
            10. Lfts. entire or nearly so.
               11. Rachis winged (Pl. XXIV, 3)...................*Rhus copallina* 163
               11. Rachis not winged.
                  12. Lfts. with a few large, gland-bearing teeth at base (Pl. XXIV, 1)
                  Ailanthus 160
                  12. Lfts. not as above.
                     13. Shrub or small tree of swamps; twigs smooth, light gray, marked with dark lenticels, fr. white or whitish; *very poisonous to touch* (Pl. XXV, 1)...............................*Rhus vernix* 163
                     13. Not as above.
                        14. Lfts. marked with little dots; shrub (Pl. XXII, 2)...*Amorpha* 157
                        14. Lfts. not as above.
                           15. Lfts. oblong-lanceolate, small (1 in. long or less).
                              16. Lfts. faintly wavy-toothed, 21–30; lvs. sometimes twice pinnate; fls. greenish, inconspicuous (Fig. 70, left, p. 152)
                              Gleditsia triacanthos, var. *inermis* 153
                              16. Lfts. entire, 5–7; low shrub to 3 ft. with yellow, 5-petalled fls. ½–1 in. in diameter.....*Potentilla fruticosa* 145
                           15. Lfts. not as above.

17. Twining vines (Fig. 73, p. 157)...........................*Wisteria* 157
17. Trees.
  18. Lfts. 7–9, usually alternate, large, to 4 in. long (Pl. XXI, 1)..*Cladrastis* 153
  18. Lfts. 7–25, opposite, smaller, not over 2 in. long.
    19. Lfts. blunt and mucronate or brts. glandular-viscid; often with stipular thorns (Pl. XXII, 1, 3).......................*Robinia* 157
    19. Lfts. acute, glaucous and appressed-pubescent below; brts. without stipular spines, green (Fig. 71, p. 155).................*Sophora* 153
10. Lfts. distinctly toothed or lobed.
  20. Sap milky (Pl. XXV, 3 and XXIV, 2).....*Rhus typhina* and *glabra* 161, 163
  20. Sap not milky.
    21. Pith chambered (Pl. VI, 4, 5)...........................*Juglans* 83
    21. Pith not chambered.
      22. Lfts. 5–11, large, 3–7 in. long (Pl. VII)................*Carya* 85
      22. Lfts. 7–17, averaging smaller, ½–4 in. long.
        23. Lvs. pinnate or sometimes partly bi-pinnate, irregularly toothed, cleft or lobed, to 14 in. long; buds hemispherical, only 2 scales showing; no true terminal bud; fr. bladdery; rather rare, cult. small tree......................................*Koelreuteria* 176
        23. Lvs. and buds not as above.
      24. Lfts. doubly serrate; pith brown; brts. dying back in winter; shrub
          Sorbaria 134
      24. Lfts. simply serrate; pith white (lvs. sometimes compound only at base, the upper part or even the whole, only lobed) terminal bud present, large. (Fig. 56, p. 137)..*Sorbus* 135
8. Shrubs or small trees; lvs. twice pinnate (cf. Fig. 92, p. 192)
  25. Lfts. very small and numerous, of 40–60 lfts. to a single pinna, about ¼ in. long and ciliate; fls. (in the hardier var. *rosea*) bright pink........*Albizia* 151
  25. Lfts. larger.
  26. Lfts. lobed.
    27. Lfts. irregulary toothed, cleft or lobed; lvs. to 14 in. long; buds hemispherical, only 2 scales showing; rather rare, cult. small tree
      Koelreuteria 176
    27. Lfts. 3–5 lobed, ovate to ovate-oblong, rarely entire; small, cult. ornamental shrub...*Paeonia* 118
  26. Lfts. entire or slightly wavy-toothed; large trees.
    28. Lfts. oblong-lanceolate, with slightly wavy-toothed margins; lvs. sometimes only once pinnate (Fig. 70, p. 152).................*Gleditsia* 153
    28. Lfts. ovate or elliptical, entire (Pl. XXI, 2)..........*Gymnocladus* 153
2. **Stems and/or lvs. with prickles or thorns.**
  29. Lvs. palmately compound, of 5–7 lfts.; cult. shrub (Pl. XXIX, 4)
    Acanthopanax seiboldianus 188
29. Lvs. pinnately compound.
  30. Lvs. once pinnate.
    31. Stems with comparatively few stout thorns (sometimes in *Gleditsia* numerous on trunk).
      32. Thorns in pairs at base of lvs.
        33. Lfts. entire or slightly toothed, showing translucent dots when held to light; rachis often prickly; shrub (Pl. XXIII, 2)...*Xanthoxylum* 159
        33. Lfts. entire, not dotted as above.

37. Lvs. pinnately veined.
  51. Twining; lvs. sometimes undivided (Pl. XLI, 1)..............*Solanum* 225
  51. Not twining.
    52. Stems with axillary thorns.............................*Crataegus* 135
    52. Stems without axillary thorns.
      53. Lvs. 4–6 lobed, squarish at apex; a stipule scar encircling brt. above
        each petiole (Fig. 43, p. 124)......................*Liriodendron* 123
      53. Lvs. 2-many-lobed, pointed or rounded at apex.
        54. Lvs. aromatic.
          55. Lvs. narrow, regularly pinnately lobed (Pl. VI, 3)
            Comptonia peregrina 83
          55. Lvs. broad, irregularly palmately lobed or undivided (Pl. XIX, 2)
            Sassafras 126
        54. Lvs. not aromatic.
          56. Lvs. and brts. with many, small, dark resin glands.
            Betula pendula var. *gracilis* 91
          56. Lvs. and brts. without resin glands.
            57. Pith angled or lobed; lvs. occasionally undivided (Pls. XIII and
              XIV).......................................*Quercus* 99
            57. Pith round, i.e. circular in cross section.
              58. Lf. lobes entire, sharp; buds long, slender, sharp-pointed.
                Fagus sylvatica var. *incisa* 98
              58. Lf. lobes toothed; buds shorter and broader and blunt.
                59. Lvs. very unsymmetrical at base, lobed only at the broad
                  apex or not at all (Pl. XVI, 1)............*Ulmus glabra* 112
                59. Lvs. symmetrical at base, sometimes with 1–4 pairs of lfts.
                  at base.............................*Sorbus hybrida* 136
36. **Lvs. mostly not lobed.**
  60. Lvs. parallel veined (fan-shaped, sometimes lobed) (Fig. 3, p. 50).....*Ginkgo* 50
  60. **Lvs. palmately veined.**
    61. Climbing vines.
      62. With tendrils.
        63. Mostly glabrous; stems usually prickly (Fig. 24, p. 73).......*Smilax* 73
        63. More or less pubescent or woolly; not prickly (lvs. sometimes lobed)
          (Fig. 83, p. 179)...*Vitis* 177
      62. Without tendrils (lvs. sometimes lobed) (Fig. 40, p. 120).*Menispermum* 119
    61. Not climbing.
      64. Sap milky; lvs. often irregularly lobed.
        65. Lvs. often opposite; bark of medium sized brs. gray, with conspicuous
          orange lenticels (Pl. XVII, 3).......................*Broussonetia* 114
        65. Lvs. all alternate; bark of medium brs. yellowish or brownish, with
          yellowish lenticels (Pl. XVII, 1 and 2).....................*Morus* 116
      64. Sap not milky; lvs. not lobed.
        66. Lvs. somewhat fleshy, the lower opposite; shrub of salt marshes (Fig.
          116, p. 242)...*Iva* 242
        66. Not as above.
          67. Lvs. toothed.
            68. Lvs. with tufts of hairs in axils of veins beneath, or whole lower
              surface tomentose (Pl. XXIX, 1 and 2).................*Tilia* 179
           68. Lvs. smooth or slightly hairy beneath.
              69. Lvs. very unsymmetrical at base, long- and slender-pointed at
                tip, prominently toothed; tree (Pl. XVI, 3)...........*Celtis* 114
             69. Lvs. symmetrical at base, acute but not long-pointed at tip,
                very shallowly toothed; shrub...................*Ceanothus* 176
        67. Lvs. entire, heart-shaped or nearly circular (Pl. XXI, 3)......*Cercis* 152

60. **Lvs. pinnately veined.**
  70. Lvs. sprinkled, at least on lower surface, with resin dots.
    71. Lvs. somewhat toothed toward tip, aromatic; pith green (Pl. VI, 1 and 2)
                                                              Myrica 83
    71. Lvs. entire or serrulate, not aromatic; pith small, white (Pl. XXXVII, 1, 2,
       and 3)..*Gaylussacia* 207
  70. Lvs. not as above.
    72. Lvs. and brts. dotted with silvery, and also sometimes, brown scales (Fig.
       88, p. 186)..*Elaeagnus* 186
    72. Lvs. not as above.
      73. Stipule scar or scars nearly or quite encircling brt. at each petiole base.
        74. Stipule scar completely encircling brt.; brts. aromatic (Pl. XVIII)
                                                   Magnolia 120
        74. Stipule scars not quite meeting around brt.; brts. not aromatic (Pl.
         XII, 4 and 7)..*Fagus* 96
      73. Stipule scars not as above.
        75. Brts. aromatic when broken.
          76. Lvs. entire.
            77. Lvs. obovate or elliptic, evenly distributed along brts.; often
               several buds in axil of a lf. (Pl. XIX, 1)..............*Lindera* 126
            77. Lvs. ovate or elliptic or irregularly lobed, very unevenly distrib-
               uted along brts.; usually only one bud in axil of each lf. (Pl. XIX,
               2)..*Sassafras* 126
          76. Lvs. toothed.
            78. Petioles glandular; brts. with cherry- or almond-like taste or odor
               (Pl. XX)....................................*Prunus* spp. 146
            78. Petioles not glandular; brts. with wintergreen taste and odor
               (Pl. VIII, 1 and 2)..........*Betula lenta* and *B. alleghaniensis* 89
        75. Brs. and brts. not aromatic when broken.
          79. Sap milky; stems sometimes with thorns; tree (Pl. XVII, 4).*Maclura* 114
          79. Sap not milky.
            80. Brs. and brts. with thorns.
              81. Brts. with lengthwise lines starting down from nodes, light
                 silvery gray; pith and wood greenish; thorns either at nodes or
                 ending short brts. or both (Fig. 110, p. 225)..........*Lycium* 227
              81. Brts. not as above.
                82. Thorns occasional, ending some of short brts.; early-flowering,
                   rosaceous shrub (Fig. 61, p. 140)..............*Chaenomeles* 141
                82. Thorns regularly occurring at nodes.
                  83. Thorns slender, replacing lvs. of main shoots, simple, or
                   3-divided (Fig. 39, p. 119)....................*Berberis* 119
                  83. Thorns stout, simple, axillary.................*Crataegus* 135
            80. Brs. and brts. without thorns.
              84. Twining.
                85. Pith pale brown and chambered (Fig. 86, p. 182)..*Actinidia* 181
                85. Pith not brown, continuous.
                  86. Lvs. toothed, ovate (Pl. XXVII, 3)............*Celastrus* 170
                  86. Lvs. entire, kidney-shaped (Fig. 36, p. 117)....*Aristolochia* 117
                  86. Lvs. entire, or pinnately or irregularly lobed at base, or
                     even compound (Pl. XLI, 1).................*Solanum* 225
             84. Not twining.

87. Lvs. very unsymmetrical at base.
  88. Lvs. coarsely wavy-toothed (Pl. XI, 3)......................*Hamamelis* 129
  88. Lvs. sharply serrate.
    89. Lvs. ovate, mostly doubly serrate, often pubescent beneath but never
      white (Pls. XV and XVI)....................................*Ulmus* 110
    89. Lvs. heart-shaped, simply serrate, or, if doubly serrate then white-tomen-
      tose beneath (Pl. XXIX, 1 and 2, and Fig. 84, p. 180).............*Tilia* 179
87. Lvs. more or less symmetrical at base.
  90. Lvs. mostly with elongated dark glands along upper side of midrib (Fig. 57,
    p. 138)..*Aronia* 136
  90. Lvs. not as above.
    91. Bark of twigs bitter.
      92. Lvs. about as broad as long, except in *P. balsamifera.* (See key to
        Populus, p. 79) (Pls. IV and V)...........................*Populus* 79
      92. Lvs. distinctly longer than broad.
        93. Lvs. entire or nearly so.
          94. Lvs. not at all toothed; elliptic or ovate, glaucous and pubescent
            beneath (but not tomentose), without stipules; fr. fleshy; buds
            with several scales (Pl. XXXVII, 4)......: *Vaccinium stamineum* 210
          94. Lvs. usually at least sparingly toothed; lanceolate or elliptic, tomen-
            tose at least when young, sometimes glaucous, stipules sometimes
            persistent; fr. dry; buds with a single cap-like scale (Pl. III and
            Figs. 25–28, pp. 75, 78)................................*Salix* 74
        93. Lvs. distinctly toothed.
          95. Buds with a single cap-like scale (Pl. III and Figs. 25–28, pp. 75,
            78)..*Salix* 74
          95. Buds with more than one scale showing.
            96. Lvs. with 2 or more glands on petiole or at base of blade; brts.
            with "cherry" odor (Pl. XX)....................*Prunus* spp, 146
            96. Not as above; buds and brts. pubescent (Fig. 60, p. 139)....*Malus* 140
    91. Bark of twigs not bitter.
      97. Lvs. entire or nearly so (2nd 97, p. 26).
        98. Lvs. appressed pubescent over whole lower surface even when mature,
          brts. green, usually in tiers or platforms (Pl. XXXI, 3)
                                     Cornus alternifolia 195
        98. Not as above.
          99. Margins of lvs. usually ciliate and midribs strigose beneath; lvs.
            often all crowded together at tips of brs. (scattered on more vigorous
            shoots) and brts. clustered at end of previous year's growth (Fig.
            95, p. 199 and Pl. XXXIII, 1 and 2)..Azalea group of *Rhododendron* 198
         99. Lvs. not both ciliate on margin and strigose on midrib; lvs. and brts.
            more scattered.
           100. Shrubs; brts. slender; lvs. averaging small (1–4 in. long).
            101. Buds hidden by base of petiole; brts. tough, swollen at joints
              (Pl. XXX, 4, and Fig. 87, p. 185)..................*Dirca* 185
            101. Buds and brts. not as above.
              102. Lvs. pubescent beneath.
                103. Lf.-bud scales usually sharp-pointed; fr. a berry with
                  many small seeds; (Pl. XXXVIII, 2 and 4).
                    Vaccinium myrtilloides and *corymbosum* 212
               103. Lf.-bud scales not as above.

104. Fr. a small pome ($\frac{1}{4}$ in. diam.) lvs. small ($\frac{1}{5}$–$\frac{1}{2}$ in. long) (Fig. 53, p. 135)

Cotoneaster horizontalis 135

104. Fr. a much larger pome, yellow-green and fragrant when ripe. . . *Cydonia* 141

104. Fr. a berry or drupe-like berry; buds naked; lvs. oval or obovate, slightly pubescent beneath, 1–3 in. long; berries black.

Rhamnus frangula 176

102. Lvs. glabrous beneath.

105. Lf.-bud scales usually sharp-pointed; fr. a berry with many small seeds (Pl. XXXVIII, 4) . . . . *Vaccinium corymbosum* 212

105. Lf.-bud scales not as above.

106. Lf. base cuneate, i. e., wedge-shaped.

107. Petiole $\frac{1}{2}$–1 in. long, often red; lvs. elliptic to ovate-oblong, 1$\frac{1}{2}$–2$\frac{1}{2}$ in. long, acute (mucronate or mucronulate in some spp.); fr. a 5-angled or lobed capsule (Fig. 52, p. 134) . . . . *Exochorda* 134

107. Petiole averaging shorter.

108. Lvs. elliptic to oblong, or obovate, 1–3 in. long, acute or obtuse, entire or sometimes serrulate; fr. a dry capsule (Fig. 97, p. 201 and Pl. XXXVI, 1 and 2) . . . . *Lyonia* 201

108. Lvs. oblong to oblanceolate, 1–3 in. long; fls. lilac to rosy purple; small, ornamental, early-flowering shrub (Pl. XXX, 3)

Daphne mezereum 183

106. Lf. base more or less rounded.

109. Lvs. mucronate at apex, elliptic-oblong to slightly obovate, entire or nearly so; fr. a red, long-stalked drupe; common n. (Pl. XXVI, 5) . . . . *Nemopanthus* 167

109. Lvs. rounded or often slightly emarginate at apex, oval to obovate; fr. stalks feathery, giving in late summer a "smoke" effect (Fig. 77, p. 165) . . . . *Cotinus* 165

100. Trees; brts. stouter; lvs. larger, mostly 2–4 (sometimes to 5$\frac{1}{2}$) in. long.

110. Pith angled or lobed in cross section, continuous.   (Pl. XIII, 3)

Quercus phellos and other "willow" oaks 109

110. Pith round in cross section, usually diaphragmed.

111. Winter buds sessile.

112. Lvs. obovate or elliptic, usually acute at both ends; lf. scars with 3 distinct bundle scars; buds with about 4 scales showing; frs. small, blue-black, 1-seeded drupes (Fig. 89, p. 187) . . . . *Nyssa* 187

112. Lvs. ovate or elliptic, often rounded at base; buds with 2 overlapping scales; frs. about 1 in. in diam. (except in cult. vars.), yellowish (Fig. 103, p. 213) . . . . *Diospyros* 212

111. Buds stalked, brown-tomentose; lvs. ovate-oblong to -lanceolate, pointed, 6–12 in. long; fls. purple, parts in 3's; fr. like a small banana, sweet and edible; small trees . . . . *Asimina* 126

97. **Lvs. distinctly toothed or notched.**

113. Buds stalked.   (See footnote, p. 32)

114. Buds glabrous; bud scales (2 or 3) valvate; cone-like, woody, fruiting catkins present on mature plants all summer (Pl. X, 1 and 2).

Alnus rugosa and *serrulata*   92

114. Buds pubescent; bud scales not valvate, 1 scale partly enfolding globular fl. bud (Fig. 48a, p. 130)..................................*Fothergilla* 130
113. Buds sessile.
  115. Brts. marked with lengthwise lines starting from nodes (lines sometimes obscured by dense tomentum).
    116. Lvs. coarsely toothed.
      117. Lvs. large, 5–10 in. long, with slender-pointed teeth; stems usually growing from stumps of dead trees (Pl. XII, 1, 2, and 3)
            Castanea dentata 98
      117. Lvs. smaller, ¾–3 in. long, with blunt-pointed teeth; shrub of salt marshes (Fig. 115, p. 241)............................*Baccharis* 242
    116. Lvs. finely toothed.
      118. Tiny dark stipules, or their scars present; frs. fleshy; lvs. usually pubescent, at least on veins beneath (Pl. XXVI, 3).......*Ilex verticillata* 167
      118. Stipules lacking; frs. dry; lvs. densely tomentose or entirely glabrous beneath (e. g., Fig. 51, p. 133)...........................*Spiraea* 133
  115. Brts. not marked with lengthwise lines.
    119. Pith chambered or diaphragmed; fls. white; ornamentals.
      120. Shrub; fls. in axillary clusters; lvs. obovate to elliptic, 1–3 in. long, sharply serrate; pith chambered; buds conical, at right angles to brts., with about 4 scales; fr. orange...................*Symplocos tinctoria* 213
      120. Small trees.
        121. Fls. bell-shaped, white, pendent in long series from brts. in spring; fr. 4-winged; bark of young brts. quickly shredding (Fig. 105, p. 216)
            Halesia 215
        121. Fls. with 2 conspicuous white bracts, the larger 6 in. long; lvs. heart-shaped, 3–5 in long, the teeth ending in bristle points; pith with firmer diaphragms at intervals; rare, cult. tree (Fig. 90, p. 188)
            Davidia 188
    119. Pith solid, not chambered or diaphragmed.
    122. Lvs. doubly and sharply serrate.
      123. Staminate catkins present and conspicuous in fall and winter; shrubs or trees.
        124. Shrubs.
          125. Staminate catkins gray; lf. teeth fairly large (Pl. XI, 1 and 2)
              Corylus 96
          125. Staminate catkins red-brown; lf. teeth smaller (Pl. X, 3)
              Alnus crispa 92
        124. Trees.
          126. Bark marked with transverse, elongated lenticels even on brts. of 5 or 6 yrs. (Pls. VIII, and IX, 1 and 3)............*Betula* 87
          126. Bark not as above, scaly, in long, irregular, more or less parallel strips; staminate catkins usually in 3's; buds with greenish tinge, not 4-sided (Pl. X, 5).................... ....*Ostrya* 94
      123. Neither staminate nor pistillate catkins evident in fall and winter; trees.
        127. Bark scaly.
          128. Petioles very short, to ⅙ in.; fl. buds spherical; brts. gray or gray-brown; lvs. sometimes simply serrate, usually each large tooth with 1 or 2 small teeth (Pl. XV, 4)............*Ulmus pumila* 114
          128. Petioles longer, ½–1 in.; fl. buds not spherical; brts. brownish; lvs. as above. (Fig. 60, p. 139)...............*Malus pumila* 140

127. Bark smooth, steel blue, fluted; buds pointed (Pl. IX, 4, X, 4, and Fig. 31,
p. 94)...*Carpinus* 94
122. **Lvs. simply serrate** (see also *Ulmus pumila* and *Malus pumila* above).
129. **Lvs. averaging large** (3–10 in.).
130. Lvs. pubescent beneath, at least when young.
131. Pubescence stellate.
132. Lvs. rounded to broad-obovate, 3–8 in. long, usually finely toothed
above the middle, rarely 3-parted at apex; pith rounded, greenish;
rare, cult. shrub...............................*Styrax obassia* 215
132. Lvs. elliptic-oblong, 3–6 in. long, with stellate pubescence beneath, at
least when young; pith star-shaped, white; rare, cult. trees.
133. Lvs. glandular beneath; brts. red-brown, fairly glabrous in age.
Castanea crenata 99
133. Lvs. not glandular beneath; brts. buff-colored, downy.
Castanea mollissima 99
131. Pubescence simple; lvs. elliptic to oblong-lanceolate, red in. fall, 3–8 in.
long; small tree, native in s. part of our area, often cult. n. (Fig. 98, p.
204).......................................*Oxydendrum arboreum* 206
130. Lvs. glabrous or nearly so.
134. Lvs. ovate-oblong, 3–7 in. long, with fine, bristle-like teeth, sparingly
pubescent sometimes on veins beneath; brts. somewhat pubescent;
rarely cult. small tree; pith rounded (Fig. 106, p. 217).....*Pterostyrax* 215
134. Lvs. oblong-lanceolate, acuminate, 4–10 in. long; brts. glabrous; pith
star-shaped in section (Pl. XII, 1, 2, and 3).........*Castanea dentata* 98
129. **Lvs. averaging smaller** (1–4 in.).
135. Lvs. with remote small teeth, elliptic to elliptic-oblong, acuminate; buds
superposed, with stellate pubescence; pith green (Fig. 104, p. 215)
Styrax japonica 214
135. Lvs. finely serrulate.
136. Fr. fleshy.
137. Fr. an edible berry, glaucous-blue, with many small seeds; brts. and
lvs. glabrous (Pl. XXXVIII, 1, 2, 3).........*Vaccinium vacillans,*
myrtilloides, and *angustifolium* var. *laevifolium* 210
137. Fr. an inedible, 1-seeded, bright blue drupe, about ⅓ in. long; brts.
pubescent; lvs. elliptic to oblong-obovate, usually pubescent beneath
(Fig. 103a, p. 214).......................*Symplocos paniculata* 213
136. Fr. a dry, globular capsule.
138. Fl. buds conspicuous in bracted racemes in fall and winter; styles
persistent on fr.; lvs. elliptic to oval-lanceolate, 1–3 in. long; native
shrub of pond shores and moist thickets (Pl. XXXVI, 4 c and d)
Leucothoë racemosa 206
138. Fl. buds not as above.
139. Lvs. obovate, sometimes entire, 1–3 in. long; low native shrub
(Fig. 97, p. 201)...........................*Lyonia ligustrina* 204
139. Lvs. elliptic to oblong-lanceolate, 3–8 in. long. Small tree native
in s. part of our area, often cult. n. (Fig. 98, p. 204)
Oxydendrum arboreum 206
135. Lvs. with blunt teeth.
140. Rare, cult. tree; fl. in Aug. and Sept. (Fig. 34, p. 112)..*Ulmus parvifolia* 114
140. Shrub or very small tree, mainly of northern bogs (Pl. IX, 2)
Betula pumila 91

135. Lvs. with sharp either large or small teeth.
  141. Lvs. with large teeth.
    142. Lvs. with short petioles ($\frac{1}{10}$–$\frac{1}{5}$ in.), much longer than broad, ovate to ovate- oblong, 1–4 in. long (sometimes longer) rough above, glabrous or nearly so beneath (Pl. XVI, 4)........................*Zelkova serrata* 114
    142. Lvs. with long petioles (2–5 in.), about as broad as long, ovate to heart-shaped, cordate, acuminate, glaucous beneath and glabrous except bearded in axils of veins, 3–10 in. long (Fig. 86a, p. 183)........*Idesia* 183
  141. Lvs. with sharp, smaller teeth.
    143. Fr. dry; lvs. obovate or elliptic-obovate; terminal bud rose-colored, much larger than lateral buds, with long-pointed, deciduous, outer scales (Pl. XXXIII, 3)...*Clethra* 198
    143. Fr. fleshy.
      144. Fr. a pome.
        145. Pome small (about $\frac{1}{3}$ in. diam.).
          146. Buds long and slender, pointed, green and/or with pinkish tinge, scales ciliate, often twisted; fls. in elongate clusters; native shrub or small tree (Fig. 59, p. 139)....................*Amelanchier* 139
          146. Buds ovoid, 4–6 scales without cilia showing; fls. in flat-topped clusters; rather rare, cult. shrub (Fig. 58, p. 138)........*Photinia* 138
        145. Pome larger (except in some small-fruited crabapples; in these calyx is deciduous).
          147. Buds and brts. pubescent; lvs. elliptic to oval, 2–4 in. long, usually somewhat pubescent beneath (Fig. 60, p. 139)............*Malus* 140
          147. Buds and brts. glabrous; lvs. similar to last in size and shape, but usually glabrous beneath, at least when mature; brts. often tending to end in thorns (Fig. 62, p. 141)......................*Pyrus* 141
      144. Fr. a drupe, or berry-like drupe.
        148. Stipules or their remnants present; fr. a red, berry-like drupe; native shrub...*Ilex* 165
        148. Stipules absent; fr. a bright blue drupe; cult. shrub or small tree (Fig. 103a, p. 214).......................*Symplocos paniculata* 214

WINTER KEY

To be used in the leafless season.

I. Gymnosperms

For this part of the key, use "Gymnosperms" in the Summer Key, pp. 13, 14.

II. Angiosperms

Seeds in an ovary. Broad-lvd. woody plants.

A. Buds Opposite or Whorled. Lvs. *Evergreen.*

For this part use "A" of the Summer Key, p. 14.

B. Buds Opposite or Whorled. Lvs. *Deciduous.*

1. Brts. and buds with numerous silvery and brown scales; shrub........*Shephérdia* 186
1. Brts. and buds not as above.
   2. Parasitic on living bark of members of the Pine Family; very dwarf ($\frac{1}{8}$- about 1 in. high), lvs. connate and scale-like..............*Arceuthòbium pusíllum* 117
   2. Not parasitic.
      3. Sap milky (best seen at spring season; in winter distinct only at above-freezing temperatures).
         4. Buds with 2 scales, the outer striate, the inner bag-shaped, pith with green diaphragm at nodes (Pl. XVII, 3), buds sometimes alternate.
         Broussonètia 114
         4. Buds with at least 3 or 4 pairs of scales showing.
            5. Buds large, reddish, sometimes intermixed with green, glabrous, juice very milky (Pl. XXVIII, 8).....................*Ácer platanoìdes* 173
            5. Buds smaller, grayish, somewhat pubescent; sap only slightly milky (Pl. XXVIII, 2)...............................*Ácer campéstre* 174
      3. Sap not milky.
         6. Stems climbing or twining.
            7. Climbing by aerial holdfast roots.
               8. Holdfast roots usually in double bands below nodes; lf. scars shield-shaped (Fig. 111, p. 229)...........................*Campsis* 227
               8. Holdfast roots irregularly developed; lf. scars crescent-shaped.
               Hydràngea petiolàris 128
            7. Aerial holdfast roots absent.
               9. Climbing by twining of stem around support or sometimes creeping and rooting in soil; stems not ridged (Pl. XLV, 4)..*Lonícera japónica* 240
               9. Climbing by twining of petioles which are persistent in winter; stems strongly ridged.....................................*Clématis* 118
         6. Stems not climbing or twining.
            10. Stems either hollow or with chambered pith.

11. Each pair of lf. scars connected laterally by a line or ridge; stems hollow only. Outer bark shreddy on older brs.; wood of brts. green.

    12. Fr. dry, kettle-shaped, persistent; brts. brown or red brown (Fig. 47, p. 127)

    12. Fr. a berry.

        13. Bundle scar 1; berries white, usually in irregular clusters, buds sometimes collateral.

        13. Bundle scars 2; berries red, rarely yellow, in pairs; buds often superposed

11. Each pair of lf. scars not connected laterally as above.

    14. Trees; buds more or less buried in bark above lf. scar (Pl. XLII).

    14. Shrubs; buds evident and scaly, often multiple on flowering stems; brts. more or less 4-sided (Pl. XL, 1)

10. Stems not hollow nor with chambered pith.

    15. Buds either partly or entirely covered by lf. scar, i. e., underneath the bark of lf. scar, or more or less buried in bark of brt. above lf. scar.

        16. Buds partly or entirely covered by lf. scar, capsules of originally top-shaped fr. splitting into 4 parts (Figs. 45 and 46, p. 126)

        16. Buds more or less buried in bark above lf. scar.

            17. Buds entirely buried in bark; lf. scars sometimes 3 or even 4 at a node (*whorled*); end of stem ordinarily dying back (Pl. XLI, 2)

            17. Buds only sunken in bark or brt. above lf. scar.

                18. Lf. scars 3 at a node, 2 large and 1 smaller (Pl. XLI, 3)

                18. Lf. scars 2 at a node; buds superposed (Pl. XLII)

15. Buds fully exposed.

    19. Buds, at least some of them, naked.

        20. Bundle scar or scar-complex 1; fr. violet colored, berries $\frac{1}{8}$–$\frac{1}{6}$ in. diam., in clusters; rather rare cult. shrub (Fig. 109a)

        20. Bundle scars 3 (see key p. 232)

    19. Buds all scaly.

        21. Buds with only a single scale showing (in *Cercidiphyllum* part of an inner scale sometimes appears in front).

            22. Bud scales solitary, cap-like; buds often sub-opposite or sometimes alternate (Pl. III, 3)

            22. Bud scale inclosing several interior scales; buds borne on little shoulders of the stem (Fig. 41, p. 122)

        21. Buds with more than 1 scale showing.

            23. Brts. aromatic or fragrant.

                24. Buds borne on little shoulders of the brt. and partly covered by lf. base (Fig. 44, p. 124)

                24. Buds not as above; small, hemispheric, superposed; brts. and pith more or less 4-sided; old fl. clusters persistent

            23. Brts. not markedly aromatic or fragrant.

                25. Inner living bark bright yellow (Pl. XXIII, 1)

                25. Inner bark not as above.

                    26. Terminal bud present and fairly large, or at least distinct, resulting in a monopodial br. system—*Monopodium*. (2nd 26, p. 33).

27. Lf. scars with only 1 bundle scar (appearing as a dot, a short cresecent or a close group of dots).

  28. Brts. usually 4-lined, sometimes winged, green or greenish; pith greenish; fr. pink with scarlet interior (Pl. XXVII, 4)....................*Euónymus* 169

  28. Brts. not as above.

    29. Brts. slender, usually less than ⅛ in. diam.; lenticels not conspicuous; 5 to 6 pairs of bud scales showing; fr. a black or bluish-black berry-like drupe; shrub much used for hedges (Pl. XL, 3)....................*Ligústrum* 221

    29. Brts. thicker, more than ⅛ in. diam.; 3 or 4 pairs of keeled sharp-pointed bud scales showing; fr. a one-seeded, dark blue drupe; small tree (Pl. XL, 4) (Fig. 109, p. 223)....................................*Chionánthus* 223

27. Lf. scars with 3 bundle scars or bundle scar groups (sometimes 5 or 7 in *Hydrangea*).

  30. Buds stalked* (in *Hydrangea* this character is more evident in the larger buds).

    31. Buds with 1 pair of scales exposed; i. e., outer pair of scales valvate or nearly so.

      32. Lf. scars normal; u- or v-shaped.

        33. Brts. and buds glabrous, terminal bud large.

          34. Terminal bud ½ or more in. long, bright crimson; small tree with white striped bark (Pl. XXVIII, 4)..........*Ácer pensylvánicum* 175

          34. Terminal bud 1 in. or more long, yellow brown, swollen at base (if fl. bud) (Pl. XLIII, 3) (see also other spp.)......*Vibúrnum lentàgo* 235

        33. Brts. and buds pubescent; terminal bud shorter.......*Ácer spicàtum* 175

      32. Lf. scars raised, being basal remnants of the petiole, and therefore often hiding the stalk of the bud (Cf. Fig. 94, p. 194 and Pl. XXXI and XXXII)....................*Córnus* (except *C. alternifòlia*) 189

    31. Buds with several pairs of scales exposed; bundle scars often projecting from surface of lf. scar; pith wide, white; brts. often twisted, pale; fl. remnants persistent....................................*Hydràngea* 128

  30. Buds sessile.

    35. Large trees.

      36. Brts. stout—more than ¼ in diam.; lf. scars shield-shaped or triangular; bundle scars in 3 compound groups, or sometimes 7 or 9 in a single series; terminal bud very large (Figs. 80, 81, p. 175)...............*Aèsculus* 176

      36. Brts. more slender, usually less than ¼ in. diam. (except on vigorous young shoots).

        37. Bundle scars usually 3 (rarely 5, 7, or 9); lf. scars u-, v-, or crescent-shaped; wood diffuse-porous (Pl. XXVIII)..................*Ácer* 172

        37. Bundle scars numerous in a crescent-shaped, elliptical, or circular series; wood ring-porous (Pl. XXXIX)..................*Fráxinus* 219

    35. Shrubs, rarely small trees.

      38. Decurrent lines extending straight downward from lf. scars; bud scales ciliate.

---

* Slicing the bud lengthwise with a razor blade will enable one to determine this character with certainty.

39. Lines dark and very hairy, especially near stem apex; remnants of fr. persistent, cylindrical (Pl. XLV, 1, b)........................*Wèigela* 238

39. Lines not prominent and not hairy; frs. more bottle-shaped, often in 3's (Pl. XLIV, 5)....................................*Diervílla lonícera* 238

38. Decurrent lines mainly absent or at least not constant; bud scales comparatively few, sometimes only 1 valvate pair exposed (Pl. XLIII and Fig. 112, p. 233)...*Vibúrnum* 231

26. Terminal bud lacking either from abortion of growing point or from dying back of part of brt. apex. Terminal bud present sometimes on short growths.

40. Terminal bud mostly lacking through abortion of terminal growing point, resulting in a falsely dichotomous br. system—*False Dichotomy.**

41. Brts., at least some of them, ending in a short thorn flanked by uppermost lateral buds or by short growths or by fruiting brts.; buds closely appressed to stem, dark brown; bud scales ciliate (Fig. 82, p. 177). *Rhámnus cathártica* 176

41. Brts. not as above.

42. Brts. stout, averaging ⅓ in. diam. or more; large trees.

43. Lf. scars nearly surrounding bud; inner bark bright yellow (Pl. XXIII, 1)..*Phellodéndron* 160

43. Lf. scars more or less circular.

44. 3 lf. scars at each node, 2 large and 1 smaller (Pl. XLI, 3)....*Catálpa* 227

44. 2 lf. scars at each node; pith chambered or brts. hollow (Pl. XLII) *Paulòwnia* 227

42. Brts. slender, usually less than ⅓ in. diam.; shrubs or small trees.

45. Buds more or less surrounded by a sheath of hairs from subtending lf. scars, or from remnants of lf. bases; small, cult., ornamental trees. (Fig. 78a, p. 173)...............*Ácer palmàtum* and *A. japónicum* 175

45. Buds without surrounding hairy sheath (except in *Viburnum dentatum* (Pl. XLIII, 2).

46. Bundle scars many in a transverse or slightly crescent-shaped line directly under bud (Pl. XL, 2)........................*Syrínga* 221

46. Bundle scars not as above.

47. Stipule scars present; older brs. speckled or striped; fr. a bladdery capsule (Pl. XXVII, 1)..........................*Staphyléa* 170

47. Stipule scars absent; buds often stalked; fr. a black or red drupe (Pl. XLIII and Fig. 112, p. 233)...................*Vibúrnum* 231

40. Terminal bud usually lacking because of dying back of part of shoot apex; remains of fls. or whole fruiting structure often persistent through most of winter; shrubs.

48. Brts. with prominent, lengthwise ridges; lower lf. scars opposite, upper alternate; subshrub of salt marches (Fig. 116, p. 242)...............*Îva* 242

48. Brts. not as above; not salt marsh plants.

49. Bundle scar 1, or 1-complex.

50. 1 pair of woolly scales exposed..........................*Búddleia* 223

50. 2 or more pairs of glabrous scales exposed.

---

* A true dichotomy, or division of the growing point at stem apex into 2 equal parts, occurs in the lower plants only.

51. Each pair of lf. scars connected by a definite line or ridge; fr. a pink or purple berry clinging to brs. in long clusters...........*Symphoricárpos orbiculàtus* 238
51. Lf. scars not connected laterally; fr. a black or dark blue drupe; shrub used for hedges..*Ligústrum* 221
49. Bundle scars 3, in *Sambucus* often 5 or 7.
 52. Brts. rather thick, pith wide, lenticels prominent (Pl. XLIV, 1, 2,)..*Sambúcus* 231
 52. Brts. more slender; pith moderate; lenticels not prominent.
  53. Lf. scars ciliate above just under bud, and line or flaps connecting them laterally also ciliate; otherwise glabrous; fr. 4 persistent shiny drupes (Fig. 64, p. 143)......................................*Rhodotỳpos* 142
  53. Lf. scars and lateral connecting line not ciliate.
   54. Dark line extending lengthwise downward from middle point between lf. scars; remains of cylindrical, dry, persistent frs. hairy (Pl. XLV, 1, b)
                          Weìgela 238
   54. No dark line extending downward as above described; remains of rounded frs. bristly; bark of flowering brts. shreddy (Pl. XLIV, 4)
                          Kolkwítzia 238

C. Buds Alternate. Lvs. *Evergreen.*

For this part use "C" of the Summer Key, p. 18.

D. Buds Alternate. Lvs. *Deciduous.*

Since this part of the key includes a large number of genera—roughly about 125—a few words about its construction may help the user. The simplest way to divide these genera would be to separate them into two groups, on the basis of whether the terminal bud is true or false (see p. 9, no. 1) a character that is fairly constant in any given species and, indeed, seems to hold in certain families. But because it is often difficult to determine with absolute certainty whether a plant *has* a true or false terminal bud I have decided to eliminate the majority of the genera by some other means before coming to this point.

The different classes that are thus eliminated are set off in bold-face type and their numbers are as follows, so that if one has a plant with "thorns," "milky juice," "diaphragmed pith," etc., he can immediately turn to that group. It is of course unavoidable that a given plant may be in several classes; e. g. it may be a "vine" and also have "diaphragmed pith"; or "thorny" with "buds buried in the bark." While I have not repeated the genera in this way in all cases, I have done so occasionally. Therefore if the plant be not found in one class, try another.

Classes with their key numbers.
 6. Vines. p. 35.
 22. Stems with thorns, prickles or bristles. p. 37.
 42. Brts. with diaphragmed or chambered pith. p. 38.
 53. Brts. with strong odor when bruised; either rank and unpleasant, or fragrant and pleasant. p. 39.
 69. Brts. with yellow inner bark. p. 41.
 71. Buds with one scale only. p. 41.
 73. Brts. with milky juice. p. 41.
 77. Buds and/or brts. with resin globules or resinous coating. p. 41.
 80. Buds stalked. p. 42.
 91. Lf. scars entirely or nearly surrounding bud. p. 43.
 98. Buds more or less buried under lf. scar or in brt. above lf. scar. p. 43.

104. Buds crowded or clustered at or near tips of brts.   p. 44.

106. Catkins naked in winter.   p. 44.

110. Buds long and slender, sharp pointed.   p. 44.

113. Brts. with distinct coloring—green; or partly green and partly red; **or dark red**; or purple (brown, yellowish- or reddish-brown, or yellow brts. **not in-cluded**).   p. 44.

120[1] True terminal bud present.   p. 45.

120[2] True terminal bud absent.   p. 47.

1. Wood homogeneous, cells mostly of one kind and without pores;* seeds naked (Fig. 15, e)..GYMNOSPERMS**

  2. Brts. very slender, tending to divide into 2 much divergent equal parts from death of terminal bud; buds very small in groups of 2 or more, buried in bark; pith very small, angled; cult. in our area (except Del.)...........*Taxodium*   67

  2. Brts. not as above; buds located at tip of short growths along main shoot (i. e. long growth), also at tip of long growth (Figs. 3 and 11, pp. 50 and 60).

    3. Fr. a cone (Cf. Fig. 15, p. 65).

      4. Short growths comparatively short (Fig. 11, a, p. 60); cones persistent for many years..*Lárix*   60

      4. Short growths much longer (Fig. 11, b, p. 60); cones disintegrating scale by scale...*Pseudolárix*   59

    3. Fr. not a cone. Seeds naked, the outer wall fleshy, simulating a drupe; short growths like those of *Larix* but thicker (Fig. 3, p. 50).........*Ginkgo*   50

1. Wood not homogeneous; either *ring-porous*, with a ring of pores extending around the brt. in the spring wood (i. e. *Quercus, Fraxinus*), or *diffuse-porous*, with the pores more or less uniformly distributed throughout the wood (e. g. *Liriodendron, Tilia*)............................DICOTYLEDONS (ANGIOSPERMS) [Except in Smilax, a MONOCOTYLEDON, where the pores appear in groups throughout the cross section of the stem (Fig. 24, p. 73).]

  5. Buds and brts. covered with silvery, and also, sometimes, brown scales (Fig. 88, p. 186)..*Elaeágnus*   186

  5. Buds and brts. not as above.

    6. **Vines. Stems twining around a support, or, more rarely** (e. g. *Rosa* and *Lycium*) **leaning or clambering.**

      7. Stems either with aerial roots or with tendrils, which act as holdfasts.

        8. Aerial roots present. *Poison to touch.* (Pl. XXV and Fig. 2, p. 7) *Rhus radicans* 165

        8. Aerial roots absent except rarely in *Parthenocissus.*

          9. A pair of tendrils (modified stipules) from the base of each lf. (Fig. 24, p. 73)...*Smilax*   73

          9. Tendrils single at the nodes, but may be branched.

            10. Tendrils terminating in flat disks; berries blue black (Pl. XXX, 1, 2).......................................*Parthenocissus*   177

            10. Tendrils without disks.

---

* Pores are seen in cross section of brt., best cut with razor blade. They would then appear as tiny holes among the other wood cells.

** These gymnosperms are included here because, having deciduous leaves, they might be mistaken for angiosperms.

11. Berries turquoise blue; pith white, without nodal diaphragms; rare, cult. sp.
Ampelopsis brevipedunculata 177

11. Berries black or dark purple; pith brown or amber, with diaphragm at each node (Fig. 83, p. 179)..*Vìtis* 177

7. Aerial roots or tendrils absent.

  12. Stems with thorns* or prickles.

    13. Thorns axillary, i. e., modified brts., occurring at nodes; stems leaning; fr. red berries often in 2's (Fig. 110, p. 225)...................*Lýcium* ± 227

    13. Only prickles present, these often longer at the nodes and scattered more or less abundantly along stems.

      14. Lf. scars flush with stem, often narrow and appearing as a dark line; prickles more or less numerous on stem; often larger and in pairs at the nodes (Fig. 67, p. 146).......................................*Ròsa* 145

      14. Lf. scars not flush with stem; petiole base much raised (cf. Fig. 65, p. 144); stems never climbing; often leaning and sometimes creeping. *Rùbus* 142

  12. Stems without thorns or prickles.

    15. Pith chambered; buds buried in swelling above projecting circular lf. scar (Fig. 86, p. 182)...*Actinídia* 181

    15. Pith solid, or with diaphragm only at nodes.

      16. Pith with a single diaphragm at nodes; scar of lf. sheaf encircling each node; buds with several loose, membranous, smooth scales.
Polýgonum aubérti 117

      16. Pith without diaphragm at nodes.

        17. Petiole base much raised; sometimes whole petiole and lfts. persistent, buds with many scales projecting at wide angle from brown-purple or gray stems (Fig. 38, p. 118)............................*Akèbia* 118

        17. Petiole base not as above.

          18. Buds several at each node.

            19. Stems green; buds surrounded on 3 sides by u- or v-shaped lf. scar...*Aristolòchia* 117

            19. Stems purplish; 1 bud above lf. scar in woolly nest; 1 or 2 others below and more or less beneath raised, circular, concave lf. scar (Fig. 40, p. 120)...................*Menispérmum canadénse* 119

          18. Buds single at each node.

            20. Buds hairy, rounded, with about 4 hairy scales; stems often 3-sided or angled (Pl. XLI, 1)..............*Solànum dulcamàra* 225

            20. At least outer bud scales smooth.

              21. Buds wart-like, projecting at right angles to stem; bud scales mucronate; lf. scars semicircular, with tiny dot-like stipule scars one each side, above (Pl. XXVII, 3)....*Celástrus scándens* 170

              21. Buds somewhat elongated, appressed to stem, nearly enveloped by 2 outer scales; lf. scars somewhat raised, often with a warty or horn-like projection on each side, especially on long shoots; bean-like pendent frs. persistent until later winter (Fig. 73, p. 157)...*Wistèria* 157

---

* As explained on p. 11, the term *thorn* is reserved for a modified organ (brt., lf., or stipules), while prickles or bristles are *emergences*, appearing in no definite place or arrangement on the stem. Thorns are often not stable characters: sometimes they appear and sometimes they are absent, as in *Robinia*. Where this condition occurs I have indicated it by a plus or minus sign following the generic name. Prickles and bristles, however, are more constant characters.

6. More or less erect trees or shrubs.
  22. **Stems with thorns, prickles or bristles.*** (2nd 22, p. 38.)
    23. Thorns present.
      24. Thorns foliar, i. e., modified lvs., and therefore subtending brts. or bud; inner bark bright yellow.
        25. Thorns single (Fig. 39, p. 119)..................*Bérberis thunbérgi* 119
        25. Thorns 3-pronged (Fig. 39, p. 119)...............*Bérberis vulgàris* 119
      24. Thorns stipular, i. e., modified stipules as shown by their occurrence in pairs at lf. base or (in *Acanthopanax*) their close connection with the lf. base.
        26. Usually 1 or 2 simple thorns (sometimes 3) produced from the center and/or sides of the lf. scar, at each node (Pl. XXIX, 4)
                                     Acanthopánax sieboldiànus 188
        26. Thorns not as above; always at sides of lf. scar.
          27. Lf. scars thin and much raised, ending in a thorn-like process at each side; stems green, with lengthwise lines; short growths prominent.
                              Caragàna arborèscens 159
          27. Lf. scars not as above.
            28. Buds buried beneath lf. scar (Pl. XXII, 3). *Robìnia pseùdoacàcia* ± 157
            28. Buds evident above lf. scar, globose, woolly and indistinctly scaly; bark somewhat aromatic (Pl. XXIII, 2)
                          Xanthóxylum americànum** 159
      24. Thorns cauline, i. e., modified brts. and therefore either subtended by lf. scars or terminating brts.
        29. Thorns subtended by lf. scars or on trunk.
          30. Thorns with 1, 2 or more prongs; buds sunken in bark at and above lf. scar; brts. swollen at nodes, zigzag (Fig. 70, p. 152)......*Gledítsia* 153
          30. Thorns without prongs.
            31. Brts. and buds clothed with silvery and sometimes brown scales (Fig. 88, p. 186)..............................*Elaeágnus* ± 186
            31. Brts. and buds not as above.
              32. Brts. with milky juice.
                33. Thorns short; ($\frac{1}{10}$–$\frac{1}{5}$ in. long)...................*Cudrània* 114
                33. Thorns longer ($\frac{1}{3}$–$\frac{3}{4}$ in. long) (Pl. XVII, 4).. *Maclùra pomífera* 114
              32. Brts. without milky juice; thorns of 2 kinds, those with definite growth and those terminating brts.***............*Crataègus* 135
        29. Thorns terminating brts.
          34. Brts. and buds clothed with silvery and sometimes brown scales (Fig. 88, p. 186)................................*Elaeágnus* ± 186
          34. Silvery scales absent.
            35. Some thorns of definite growth,*** others terminating brts.

---

* See note, p. 36.

** Whether this pair of outgrowths at each node is really of stipular origin has been questioned. I am inclined to think that their position and regular occurrence justify this conclusion. However, the true test is whether or not they are connected up, internally, at least in the young stage, with the vascular system of the lf.; if not, they are merely prickles.

*** By definite growth I mean here growth of a simple thorn to a length which is fairly constant, so much so that in *Crataegus* such thorns are usually considered a reliable diagnostic feature for each sp.; also in *Crataegus*, there are some brs. of more or less indefinite growth provided with buds, and these brs. or brts. terminate in thorns.

36. Thorns of definite growth usually numerous; buds usually rounded, glabrous; lf. scars raised; shrubs and small trees............*Crataègus* 135

36. Thorns of definite growth usually few, the majority terminating the brts.; buds more pointed, pubescent; lf. scars raised; large, spherical fl. buds in clusters often opening in mild periods in winter; shrubs, sometimes half evergreen......................*Chaenomèles lagenària* 141

35. All thorns terminating brts.; decurrent ridges from sides and center of lf. scars; buds and brts. glabrous; brts. rather coarse and heavy (Fig. 62, p. 141)..*Pỳrus commùnis* 141

23. Thorns, as such, absent; prickles or bristles present.

37. Lf. scars much raised because of remains of petiole base (cf. Fig. 65, p. 144)

38. Brts. (canes) coarse and heavy, not glaucous; prickles strong and numerous.....................................Blackberries. *Rùbus* spp. 142

38. Brts. (canes) more slender, often rooting at tip, glaucous; weak prickles and/or bristles present (Fig. 65, p. 144)......Raspberries. *Rùbus* spp.* 142

37. Lf. scars not or only a little raised.

39. Buds more or less buried under lf. scar; brts. bristly (Pl. XXI, 1)
Robínia híspida 159

39. Buds not as above.

40. Lf. scars describing the outline of a shield, reaching nearly around the thick gray brts. and showing at least 20 bundle scars....*Aràlia spinòsa* 189

40. Lf. scars not shield-shaped: crescent-shaped or almost linear; prickles conspicuous at nodes.

41. Brts. shreddy (shedding epidermis early), with decurrent ridges from the nodes; lf. scars slightly raised......Gooseberry section of *Ríbes* 128

41. Brts. not shreddy, without decurrent ridges; often bright green or red; lf. scars flush with brts., often in a narrow line; prickles abundant, especially on young shoots.........................*Rósa* 145

22. Thorns, prickles or bristles absent.

42. **Brts. with diaphragmed or chambered pith.**

43. Only 1 thin green diaphragm at each node; buds mostly covered by a single, striate scale; buds either opposite or alternate; juice milky (Pl. XVII, 3)
Broussonètia 114

43. Pith plentifully supplied either with diaphragms or chambers.

44. Brts. with true terminal buds (see p. 9, "1").

45. Terminal bud either with few scales and these not of the typical scale form, or naked, i. e., without scales.

46. Terminal bud with scales; pith chambered (Pl. VI, 4, 5)...*Jùglans*** 83

---

* This method of separating Blackberries and Raspberries is open to criticism because to the uninitiated it is difficult to distinguish between a *strong* prickle and a weak prickle; and, as a matter of fact, sometimes the difference *is* slight. The best method of differentiation is of course by the fr., which, in the Raspberries comes off the receptacle when ripe leaving a cone; while in Blackberries the receptacle comes off *with* the fr. The Raspberry stems too, in the common spp., the Red and the Black, are glaucous.

** *J. cinerea*, Butternut, has dark brown pith; *J. nigra*, Black Walnut, light brown.

46. Terminal bud naked; lateral ones, if fl. buds, rounded, stalked and brown woolly; diaphragms or chambers sometimes indistinct.........*Asímina* 126

45. All buds with many scales; in Magnolia and Liriodendron (see below) scales and rudimentary lvs. alternate throughout structure of bud.

   47. Brts. shreddy due to sloughing off of epidermis; true terminal buds at least on short brts. (Fig. 105, p. 216).......................*Halèsia* 215

   47. Brts. not shreddy.

      48. Stipule scars encircling brt. at each node (Fig. 43, p. 124).

         49. Outermost bud scales 2, valvate...................*Liriodéndron* 123

         49. Outermost bud scale apparently single (actually composed of 2 united stipules); diaphragms not always evident.........*Magnòlia* 120

      48. Stipule scars absent; bundle scars 3.

         50. Buds stout, ⅓ in. long or more, and thick; scales dark red, shining; diaphragms often extending only part way from sides; rare, cult. tree...*Davídia* 188

         50. Buds not so large, averaging ⅛ in. long; scales brown or red brown, often finely ciliate; lateral buds often absent on short growth; diaphragms extending all the way across but usually irregularly spaced; brts. at wide angles from main stem; bundle scars distinct, forming a curved row (Fig. 89, p. 187)..........................*Nýssa* 187

44. Brts. without true terminal buds.

   51. Stipule scars present; buds small, somewhat flattened, appressed close to brts.; insect galls and "witches' brooms" usually conspicuous; pith very narrow, either chambered or diaphragmed (Pl. XVI, 3)...........*Céltis* 114

   51. Stipule scars absent.

      52. Buds rather large (¼ in. long), with 2 or 3 overlapping scales; chambers or diaphragms sometimes absent, but then hollow or irregular chamber-like cavities apparent; tree; bark in thick polygonal chunks (Fig. 103, p. 213)...*Diospýros* 212

      52. Buds smaller (⅛–⅕ in. long), 4 to 6 scales showing; shrub
                                      Sýmplocos tinctòria* 213

42. Brts. without diaphragmed or chambered pith.

   53. Brts. with strong odor when bruised, either rank and unpleasant or fragrant and pleasant.** (See also Photinia and Exochorda, nos. 158² and 132²) (2nd 53, p. 41).

      54. Odor rank and unpleasant.

         55. Brts. thick and heavy, brown, tinged with green; lf. scars large, with bundle scars arranged in horseshoe outline; no true terminal bud; stipule scars absent (Pl. XXIV, 1)..............................*Ailánthus* 160

         55. Brts. more slender, dark red brown (yellow brown in *Prunus avium*); buds reddish brown; stipule scars present; taste and odor like that of almonds (Pl. XX).......................................*Prúnus**** 146

---

* In the specimens examined, *Symplocos paniculata*, the common cult. exotic sp., did not show chambered or diaphragmed pith.

** Members of the Pea or Pulse Family are omitted from this category. They usually have a leguminous odor when bruised and a taste like that of raw peas or beans.

*** Comprising both Cherries and Plums; Cherries have fr. with rounded pits and a true terminal bud; Plums have flattened pits and no true terminal bud.

54. Odor fragrant and pleasant.
  56. Brts. bright green; true terminal bud present.
    57. Internodes of very variable length; brts. widely divergent; stipule scars absent; odor strong (Pl. XIX, 2)..........................*Sássafras* 126
    57. Internodes not as above; but brts. may be widely divergent; stipule scars encircling brt.; odor not strong but plainly aromatic (Pl. XVIII, and Fig. 42, p. 123)...*Magnòlia* 120
  56. Brts. not bright green.
    58. Brts. with stipular thorns; odor very faint (Pl. XXIII, 2)
                                        Xanthóxylum americànum 159
    58. Brts. without thorns.
      59. Catkins or catkin-like buds present.
        60. Stipule scars present; catkins and brts. pubescent, with abundant resin globules; brts. slender; lf. scars semicircular, with tiny stipule scars at each side; buds round and shining; low shrubs in sandy soil (Pl. VI, 3).....................................*Comptònia peregrìna* 83
        60. Stipule scars absent.
          61. Lf. scars elliptical, lf. buds more pointed than in last; catkin-like buds cone-like; often with resinous or waxy coating; larger shrub than last, frequenting borders of ponds and swamps (Pl. VI, 2)
                                        Myrìca gàle 83
          61. Lf. scars circular; lf. buds hairy, hidden under upper flap of lf. scar (Fig. 76, p. 163)..............................*Rhus aromática* 163
      59. Catkins or catkin-like buds absent.
        62. Resin globules plentifully sprinkled on buds, especially the terminal cluster, and sparingly on brts; buds small, globular, dark red, shining; bruised twig with odor of barber's bay rum (Pl. VI, 1)
                                  Myrìca pensylvanica 83
        62. Resin globules absent.
          63. Stipule scars encircling brts.; true terminal bud present.
            64. Terminal bud flattened like duck's bill (Fig. 43, p. 124)
                                  Liriodéndron 123
            64. Terminal bud not as above (Pl. XVIII and Fig. 42, p. 123)
                                  Magnòlia 120
          63. Stipule scars, if present, not encircling brts.
            65. True terminal bud present; pith brown.
              66. Terminal bud large, prominent, dome-shaped; lf. scars large, heart- or shield-shaped (Pl. VII, 2).........*Cárya tomentòsa** 85
              66. Terminal bud small, often surrounded by tardily dehiscing lf. bases; lf. scars crescent-shaped; sap aromatic and gummy (Fig. 77, p. 165)..............................*Cótinus coggýgria* 165
            65. True terminal bud absent.
              67. Buds of 2 kinds: stalked, rounded fl. buds, and sessile, smaller, narrowly ovoid lf. buds; brts. slender, greenish brown, with spicy odor (Pl. XIX, 1)....................*Líndera benzòin* 126
              67. Buds of only one kind.

---

  *All spp. of Hickory possess some fragrance in lvs. and young bark but in this one the odor is striking.

68. Brts. with wintergreen taste and odor; buds sharp-pointed, slender; many short growths along main brts. (Pl. VIII, 1, 2)

Bétula lénta and *alleghaniénsis** 89

68. Brts. without wintergreen taste or odor; buds pubescent, surrounded on 3 sides by lf. scar; odor of bruised bark suggesting that of turpentine (Pl. XXIII, 3)..*Ptèlea trifoliàta* 159

53. Brts. without strong odor when bruised.

69. **Brts. with yellow inner bark.**

70. Low, little-branched shrub, with bright-yellow inner bark; roots yellow; lf. scars rather narrow, more than half encircling brt...........*Xanthorhìza* 118

70. Medium sized tree; brts. with greenish-yellow inner bark; lf. scars almost circular (Fig. 86a, p. 183)...................................*Idèsia* 183

69. Brts. not as above.

71. **Buds with only one scale,** which covers bud like a cap or inverted bag (sometimes flattened in *Salix*). (The *Magnolia* bud is also covered by a single scale, but throughout the lf. bud, scales and rudimentary lvs. alternate).

72. Petiole base covering bud during growing season (Fig. 49, p. 131)

Plátanus 132

72. Petiole base not as above (Pl. III)..........................*Sàlix* 74

71. Buds with more than one scale.

73. **Brts. with milky juice.** (Best seen, when not in the growing season, in early or late winter, at above-freezing temperatures, in cross-section of brt.); true terminal bud absent.

74. Brts. especially the vigorous long growths, with single, axillary thorns, spherical buds at side of the thorns (Pl. XVII, 4)........*Maclùra* ± 114

74. Brts. without thorns.

75. Stipule scars present; small or large trees.

76. Buds often opposite or subopposite; 2 bud scales exposed, the outer and sometimes the inner striate (Pl. XXVII, 3)

Broussonètia 114

76. Buds always alternate; 3–6 scales exposed in lf. buds; brts. with yellowish hue; milky juice often scanty; fl. buds larger than lf. buds, rounded (Pl. XXVII, 1, 2).......................*Mòrus* 116

75. Stipule scars absent; shrubs, sometimes tree-like (Pls. XXIV, XXV)..*Rhus* spp. 161

73. Brts. without milky juice.

77. **Buds and/or brts. with resin globules or resinous coating.**

78. Buds and/or brts. showing, under lens, resin globules.

79. Brts. aromatic and fragrant (Pl. VI, 1, 2, 3)

Myrìca and *Comptònia* 83

(See Nos. 60¹, 61¹, 62¹ of this key, p. 40.)

79. Brts. not aromatic; fl. buds large, lf. buds small (Pl. XXXVII, 1–3)..*Gaylussàcia* 207

---

* *B. alleghaniensis* was formerly *B. lutea.* Brts. of Sweet and Yellow Birches are almost identical in taste and odor, except that the Yellow B. is milder and the brts. tend to be slightly hairy, especially in the region of the buds, and sometimes show a slightly yellowish hue.

78. Buds, except for outermost scale, encased in hardened resin; brts. ridged; pith with crenulate margin in section; shrubs of salt marshes (Fig. 115, p. 241)..*Báccharis* 242

77. Buds without resin globules or resinous coating.

   80. Buds stalked, either the lf. buds or the fl. buds or both.

      81. Some or all of the bud scales more or less flattened, straight on one edge, curved on other.

         82. True terminal bud present.

            83. Stipule scars present.

               84. Shrubs.

                  85. Buds yellowish gray; fls. yellow; common native and cult. shrubs (Pl. XI, 3)............................*Hamamèlis* 129

                  85. Buds brown, 1 scale partly enfolding globular fl. bud; fls. white; rather rare cult. shrub from s.e. U. S. (Fig. 48a, p. 130)
                                              Fothergílla 130

               84. Small tree; buds nearly black; blackish bark flaking off much as in *Cornus kousa*; rarely cult.........................*Parrótia* 129

            83. Stipule scars absent; terminal bud knife-like when a lf. bud, globose if a fl. bud; lateral buds globose; pith with diaphragms, sometimes indistinct...................................*Asìmina trilòba* 126

         82. True terminal bud absent; stipule scars absent; brts. slender, shreddy or with exfoliating bark; buds small, superposed, yellowish scurfy; pubescence stalked-stellate; pith and wood of brts. green (Fig. 104, p. 215)..*Stỳrax* 214

      81. Buds provided with typical scales.

         86. Buds with only 2 or 3 valvate scales.

            87. Shrubs or small trees; pith white; naked catkins present in winter and also remnants of old cone-like frs. (Pl. X, 1, 2,)
                        Álnus (except *A. crispa*, var. *mollis*)   91

            87. Large forest trees; pith brown; buds sulphur yellow, often superposed (Pl. VII, 4)...........................*Cárya cordifórmis*   87

         86. Buds with numerous scales; or, if only 2 or 3, these not valvate.

            88. Fl. buds differentiated from lf. buds in size and shape, fl. buds stalked.

               89. Fl. b uds large ($\frac{1}{4}$ in. diam.) usually solitary, round or oval, often accompanied by small, basal, lf. buds..............*Corylópsis* 129

               89. Fl. buds smaller, sometimes solitary but more often in pairs or clusters.

                  90. Shrubs; brts. with spicy fragrance; fl. buds spherical (Pl. XIX, 1)..*Lindera* 126

                  90. Small trees; fl. buds rounder than lf. buds, with finely white-ciliate bud scales becoming progressively shorter toward the base of bud, concealing the stalk; fl. buds often in clusters on the old wood (on short growths) (Pl. XXI, 3).........*Cércis* 152

            88. Fl. buds not differentiated from lf. buds; of about the same size and shape although the latter may be smaller; bud scales loose; short growths conspicuous, closely crowded, with many sharp-pointed cylindrical buds about $\frac{1}{8}$ in long (spp. with prickles belong to the Gooseberry group).....................Currant section of *Rìbes* 128

   80. None of the buds stalked.

91. **Lf. scars entirely or nearly surrounding bud.**
  92. Lf. scars entirely surrounding bud; bud scale single, cap-like; large trees with either white or yellowish inner bark (Fig. 49, p. 131)..........*Plátanus* 132
  92. Lf. scars nearly surrounding bud or at least on 3 sides of bud.
    93. Lf. scars nearly surrounding bud.
      94. Juice milky; pith light brown (Pls. XXIV, 2 and XXV, 3)
            Rhus glàbra and *typhìna* 163, 161
      94. Juice not milky.
        95. Shrubs with brts. very pliable and tough (Pl. XXX, 4).........*Dírca* 185
        95. Trees; brts. not as above; buds superposed so close together as to appear as one (Pl. XXI, 1)...........................*Cladràstis* 153
    93. Lf. scars on 3 sides of bud only.
      96. Stipule scars present at upper edges of lf. scar; brts. dark green, puberulent; with leguminous odor and taste when bruised (Fig. 71, p. 155)
            Sophòra japónica 153
      96. Stipule scars absent.
        97. Bundle scars 3; small tree; brts. olive green or brownish, glabrous, with faint odor of turpentine when bruised (Pl. XXIII, 3)..........*Ptèlea* 159
        97. Bundle scars more numerous (5–9);* shrub or low tree; brts. downy, reddish lenticels prominent (Pl. XXIV, 3)..........*Rhus copallìna* 163
91. Lf. scars not as above.
  98. **Buds more or less buried under lf. scar, and/or in tissues of brt. above lf. scar.**
    99. All buds more or less buried under lf. scar.
      100. Buds not evident.
        101. Buds buried under bark of lf. scar; numerous circular pads marking dehiscence of fr. or of fls.; hair-like stipules persistent; light-gray lenticels conspicuous (Fig. 85, p. 181)...........*Hibíscus syrìacus* 181
        101. Buds buried under upper flap of lf. scar; circular pads and stipule scars or stipules absent; buds with long yellowish hairs (Fig. 76, p. 163)
            Rhus aromática 163
      100. Buds evident, superposed, partly hidden beneath a membrane left after fall of lf. which shows, on cracking open, fine ciliated edges; pith brown or sometimes white; stipular thorns may or may not be present (Pl. XXII, 3)...................................*Robínia pseùdoacàcia* 157
    99. Buds not as above.
      102. Some buds under lf. scar, some exposed above lf. scar or in tissue above lf. scar (see also *Amorpha fruticosa* no. 139[1], p. 47).
        103. Brts. with greenish hue; pith green, sharply angled; buds superposed, at least 2 very close together, upper large and exposed, lower small and partly hidden by lf. scar..........................*Albízia* 151
        103. Brts. without greenish hue, swollen at nodes, noticeably zig-zag; thorns, when present, axillary and usually branched; lenticels on 2 yr. brts. prominent and wart-like; pith white or pinkish (Fig. 70, p. 152)
            Gledítsia triacánthos 153
      102. All buds buried in bark above lf. scar; brts. stout, grayish; pith salmon colored (Pl. XXI, 2)..........................*Gymnócladus dioìcus* 153
  98. Buds not as above.

---

* If the bundle scars are not clearly defined, a cut with a razor blade just beneath the surface of the lf. scar and in a plane parallel to it will bring them out clearly.

104. **Buds crowded or clustered at or near upper ends of brts.**

    105. Fl. buds much larger than lf. buds, and central and single in the terminal group; bud scales mucronate (Fig. 95, p. 199). *Azàlea* spp. of *Rhododéndron** 198

    105. No such clear distinction between fl. and lf. buds (Pls. XIII and XIV)

        Quércus 99

104. **Buds not crowded or clustered at or near upper ends of brts.**

    106. **Trees and shrubs with naked catkins in winter.**

        107. Both male and female catkins exposed; catkins dark red brown, winter buds stalked (Pl. X, 1, 2)................*Álnus* (except *A. crispa*) 91

        107. Only male catkins exposed during winter.

            108. Shrubs; catkins gray, often curved or curled due to insect injury; buds ovoid, blunt, with 4–6 scales showing (Pl. XI, 1, 2)...........*Córylus*** 96

            108. Trees (except *Betula pumila*, q.v. Pl. IX, 2).

                109. Catkins comparatively short, gray, slightly more than ½ in. long, often in 3's; buds rather plump, with clearly striate scales; outer bark of mature trunks in ragged, more or less vertical strips; pith pale; usually a small tree (Pl. X, 5)..............*Óstrya virginiana* 94

                109. Catkins longer, brown, not so conspicuously in 3's; buds varying in the different spp., but scales often with lengthwise lines (not clearly striate as in *Ostrya*); lenticels conspicuous, extended to transverse lines on older brts.; small or large trees; bark either in large plates, or scaly, papery or curly (Pls. VIII and IX, 1, 2, 3).....*Bétula* 87 (See also *Myrica* and *Comptonia* (Pl. VI), and *Rhus aromatica* (Fig. 76, p. 163).

    106. **Trees and shrubs with naked catkins absent.**

        110. **Buds long and slender, sharp-pointed.**

            111. Buds at least ¾ in. long; stipule scars linear, reaching well around brt., bark of old trunks smooth, gray (Pl. XII, 4–7)...............*Fàgus* 96

            111. Buds, at least the terminal ones, shorter, but at least ⅓ in. long.

                112. Buds ⅓–½ in. long, with green or pink hues; bud scales sometimes twisted, often ciliate; shrubs or small trees with smooth bark (Fig. 59, p. 139)..*Amelánchier* 139

                112. Buds, at least the terminal ones, about ⅓ in. long, crimson; lateral buds smaller, crimson, closely appressed to slender brts.; shrubs only (Fig. 57, p. 138).................................*Arònia* 136

    110. **Buds not as above.**

        113. **Brts. with distinct coloring; green, or partly green and partly red, or dark red, or purple** (brown, yellowish- or reddish-brown or yellow brts. not included). (See also *Prunus persica* and *Quercus robur*.)

            114. Brts. bright breen.

                115. Brts. aromatic, internodes of markedly irregular length; large true terminal buds with many scales (Pl. XIX, 2)..........*Sássafras* 126

                115. Brts. not aromatic.

---

* The buds of *Rhodora* are similar, but smaller and plainly somewhat glaucous.

** In *C. americana*, brts. have, toward apex, stiff hairs or remnants thereof; in *C. cornuta* these are lacking and brts. are more slender.

116. Brts. long and slender, zigzag, slightly ridged, dying back at tip; buds divergent at angle of about 45°; cult.........................*Kérria* 142

116. Brts. not as above, yellow green, low; fl. buds larger than lf. buds; fr. a berry (Pl. XXXVIII, 1) (Also other spp. of Blueberry)

Vaccínium vacíllans 212

114. Brts. darker green.

117. Large trees; brts. dark green, glabrous or slightly pubescent; buds hairy, almost hidden under lf. scars (Fig. 71, p. 155).................*Sophòra* 153

117. Shrubs or small trees.

118. Brts. hoary green near tip of long growth, otherwise dull grayish green; short growths numerous and conspicuous; stipules persistent as small knobs or longer growths at each side of lf. scar; remnants of legumes persistent...*Labúrnum* 153

118. Brts. green but not hoary; slender, terete or ribbed; lf. scars raised, very small....................................*Genísta* or *Cýtisus** 155

114. Brts. with either red or green coloring or both; large trees; buds with a large, colored scale at one side producing a hump (Pl. XXIX, 1, 2).......*Tília* 179

114. Brts. pink; low shrubs; buds, especially fl. buds, sprinkled with resinous granules (Pl. XXXVII, 1–3)............................*Gaylussàcia* 207

114. Brts. purple red above, crimson beneath; cortex rosy, especially toward upper side; small trees; buds slightly stalked; true terminal bud present; brts. in tiers (Pl. XXXI, 3)........................*Córnus alternifòlia* 195

114. Brts. dark red.

119. Brts. stout; buds large, naked, brownish hairy, superposed, often 3 close together, as in *Cladrastis*; lf. scars often nearly surrounding bud; bark of young brts. soon exfoliating....................*Stýrax obássia* 215

119. Brts. more slender; buds not as above, small, with 2 or 3 red, glabrous scales showing; lf. scars raised, semicircular; bundle scar u-shaped, compound (Fig. 98, p. 204)....................*Oxydéndrum arbòreum* 206

113. Brts. without distinct coloring.

120. **True terminal bud present.** 2nd 120, p. 47.

121. Some or all of the buds naked, i. e., without typical scales.

122. All buds naked; shrubs or small trees; brts. sprinkled with white lenticels; circular pads where fr. was borne...........*Rhámnus frángula* 176

122. Only terminal bud naked; lateral buds ovoid and rather closely appressed, 2 scales showing; bark of brts. yellow or yellow brown, sometimes mottled. (Fig. 106, p. 217)....................*Pterostýrax* 215

121. None of the buds naked.

123. Terminal bud at least twice as large as lateral buds.

124. Brts. swollen at apex, terminal bud wide but low; lateral buds small, subtended by circular or shield-shaped lf. scars. (Fig. 86a, p. 183)

Idèsia polycárpa 183

124. Brts. not as above.

125. Lf. scars large, shield- or heart-shaped.

126. Pith slender, star-shaped in section, sometimes brown (Pl. VII)

Cárya spp. 85

---

* These two genera are very closely related. *Genista tinctoria* has green striped brts. while *Cytisus scoparius* has strongly ribbed brts.

126. Pith wide, white, circular in section; *poison to touch. Handle only with paper covering* (Pl. XXV, 1)..........................*Rhus vérnix* 163

125. Lf. scars smaller.

127. Lf. scars not or only slightly raised, triangular; bundle scar one, projecting from surface of lf. scar; lateral buds often developing into short growths in their first season (Pl. XXXIII, 3).......................*Cléthra* 198

127. Lf. scars raised; bundle scars 3 or more.

128. Buds glabrous; terminal buds very large in comparison with laterals, shining, green or olive green; brts. often with corky wings; large forest trees with light gray bark on upper trunk or on limbs (Pl. XIX, 3)
Liquidámbar 129

128. Buds pubescent or hairy (outer scales glabrous in *Sorbus americana*).

129. Lf. scars crescent-shaped or linear; bundle scars 3, 5, or rarely 7; outer scales of terminal buds either glabrous and gummy, or covered with long, matted hairs; pith rounded, brownish (Fig. 56, p. 137)
Sórbus 135

129. Lf. scars linear; bundle scars 3; all bud scales more or less pubescent; long growth with lines descending from lf. scars, pubescent at nodes; pith white, somewhat angular in section and surrounded by groups of thin-walled cells; usually only short growths with true terminal buds (Fig. 60, p. 139)................................*Màlus* 140

123. Terminal bud not or only a little larger than lateral buds.

130. Buds with only 2 or 3 scales showing.

131. Brts. slender, less than $\frac{1}{16}$ in. diam., glabrous; bud scales ciliate; lf. scars triangular or crescent-shaped; stipule scars absent; bundle scar 1 (Pl. XXVI, 5).......................................*Nemopánthus* 167

131. Brts. thicker, $\frac{1}{16}$ in. or more in diam., pubescent; lf. scars semicircular; stipule scars present; bundle scars 3 or the lowest compound (Pl. X, 3)
Álnus críspa var. *móllis* 92

130. Buds with more than 3 scales showing.

132. Stipules or their scars present.

133. Stipules persistent on much raised lf. scars; small shrubs with quickly exfoliating bark; pith brown....................*Potentílla fruticosa* 145

133. Stipules not as above, but their scars present.

134. Brts. especially those of 2 yrs. or more, with shredding bark and with lines running downward from lf. scars; buds cylindrical, with about 5 more or less loose scales; stipule scars usually as tiny projections on each upper side of lf. scar. (Fig. 50, p. 133)..*Physocárpus opulifòlius* 133

134. Brts. without shredding bark, but sometimes lined downward from the nodes (Ilex).

135. Buds solitary.

136. Low shrubs; lateral buds often developing into short growths during their first yr.; circular fr. remnants persistent in winter.
Ceanòthus americànus 176

136. Trees; lateral buds not developing as above; fl. buds often much larger than lf. buds; lf. scars variable in shape in the different spp.; stipule scars narrow; pith 5-sided, green or brown; wood diffuse-porous; bark very bitter (Fig. 1, p. 3 and Pls. IV and V)................*Pópulus* 79

135. Buds superposed, a small or smaller one usually at the base of a larger one; lf. scars crescent-shaped or semicircular; bundle scar u-shaped, compound; stipule scars present as small dots, points, or spicules on each upper side of lf. scar (Pl. XXVI, 3)..*Ílex* 165

132. Stipules or their scars absent; lf. scars only a little raised; brts. reddish; odor rank when bruised (Fig. 52, p. 134)........................*Exochórda* 134

120. True terminal bud absent.

137. Brts. dying back from tip in fall and winter.

138. Brts. very long and slender; pith very small, excentric; buds small, usually multiple (collateral); lf. scars absent; base of the scale lvs. persistent; buds often replaced by circular scars of deciduous short growths.. *Támarix* 182

138. Brts. not as above; pith central, brown or yellow brown.

139. Buds superposed, 1 or 2 below the main bud, rarely opposite or subopposite; bundle scars 3, the central often projecting; stipule scars present; pith yellow brown, rarely white; brts. lined from the nodes (Pl. XXII, 2)..*Amórpha* 157

139. Buds solitary, or often small collateral buds present; lf. scars large, with 3 large, often projecting, bundle scars; buds unfolding very early in the spring (March); pith brown...............*Sorbària sorbifòlia* 134

137. Brts. not characteristically dying back in fall or winter.

140. An obvious distinction between fl. and lf. buds, either in size, shape, or position on plant.

141. Fl. buds on special brts., i.e., long (1–2 in.) bracted racemes (Pl. XXXVI, 4, c)...............................*Leucóthoë racemòsa* 206

141. Fl. buds not on special brts., but clearly distinguishable from lf. buds by their larger size.

142. Shrubs; bundle scar 1; stipule scars absent; brts. brightly colored— green, pink, red or yellow.

143. Fl. buds usually sprinkled with resinous globules; brts. pink or green; fr. a berry with 10 large seeds (Pl. XXXVII, 1, 2, 3) *Gaylussàcia* 207

143. Fl. buds without resinous globules.

144. Scales of fl. buds markedly mucronate; outer scales of lf. buds often (but not always) produced to long, sharp points, nearly valvate; fr. a berry with many small seeds (Pl. XXXVIII, 1, 2, 3, 4)..*Vaccìnium* 210

144. Scales of fl. buds not mucronate; brts. yellowish; frs. urn-shaped, dry, persistent in winter (Pl. XXXVI, 1, 2).....*Lyònia mariàna* 201

142. Trees; bundle scars 3; stipule scars present; brts. not brightly colored.

145. Bud scales in 2 ranks, buds often at one side of lf. scar; bark of trunk fissured and ridged; brts. sometimes with corky ridges (*Ulmus racemosa*) (Pl. XV, 1, 2, 3, 4; Pl. XVI, 1, 2; Fig. 34, p. 112) *Úlmus* 110

145. Bud scales in 4 ranks; buds not on one side of lf. scar; no corky ridges; bark of trunk smooth and fluted (Pl. IX, 4; Pl. X, 4) *Carpìnus* 94

140. No obvious distinction either in size, shape or position between fl. and lf. buds.

  146. Buds with not more than 2 bud scales exposed.

   147. Bundle scar 1.

    148. Pith spongy, sometimes diaphragmed; bud scales 2 (sometimes 3), greatly overlapping; trees; bark of trunk in thick polygonal chunks (Fig. 103, p. 213).........................*Diospỳros virginiàna* 212

    148. Pith not as above; shrubs.

     149. Stipule scars present, or stipules rather persistent on each side of lf. scar, leaving narrow indistinct scars after falling; 2 outer bud scales mostly somewhat parted disclosing hairy interior; fr. a small pome; common, ornamental, much branched shrubs; cult.   (Fig. 53, p. 135)
Cotoneàster 135

     149. Stipule scars absent; buds with 2 scales exposed, sometimes curved, flattened against yellowish or mottled brt.; fr. in dry clusters, persistent (Fig. 97, p. 201)......................*Lyònia ligustrìna* 204

   147. Bundle scars 3; bud scales slightly overlapping and sometimes disclosing at tip the hairy interior; lf. scars large, shield-shaped; at least lowest bundle scar compound; brts. stout, zig-zag; internodes short; lenticels prominent on older wood.....................*Koelreutèria paniculàta* 176

  146. Buds with not more than 3 bud scales exposed.

   150. Buds small ($\frac{1}{16}$ in. long or less), triangular, appressed to brts., often red, with hairy interior; lf. scars narrow, raised; bundle scars 3, simple; stipule scars small, indistinct; brts. of long growth dark red or purple, often with 3 ridges descending from lf. scars..................*Cydònia* 141

   150. Buds larger ($\frac{1}{8}$ in. or more), rather plump; stipule scars larger and distinct, unequal; bundle scars 3 or scattered or compound.

    151. Buds blunt, a much larger scale at one side, giving the whole a lopsided appearance; scales dark green or red (dark red and pubescent in *Tilia tomentosa*); pith rays distinct in cross section, with 10 × lens (Pl. XXIX, 1, 2)..*Tilia* 179

    151. Buds rather acute (blunt in *C. sativa*); bud scales not asymmetrically arranged as above; often only 2 scales exposed, brown or very dark red (*C. sativa*); pith rays not pronounced in cross section (Pl. XII, 1, 2, 3)
Castànea 98

  146. 4 or more bud scales exposed.

   152. Stipule scars present; trees.

    153. Phyllotaxy uniformly $\frac{1}{2}$.

     154. Bud scales plainly striate, in several ranks; small trees with bark of trunk exfoliating in irregular, vertical strips; catkins short, often in 3's, evident in winter (Pl. X, 5)..........................*Òstrya* 94

     154. Bud scales not markedly striate.

      155. Bud scales in 2 ranks; buds often at one side of lf. scar; bundle scars 3 or in 3 groups (Pl. XV, 1, 2; Pl. XVI, 2; Fig. 34, p. 112)...*Ûlmus** 110

      155. Bud scales in 4 ranks; buds small; brts. very slender.

_____

* *Ulmus* and *Carpinus* are included here again because in young trees or vegetative shoots there are no fl. buds.

156. Buds angled; lf. scars raised, often crater-like; bundle scars 3; small trees with bark or trunk smooth but fluted, of a steely color; fr. a small nut (Pl. IX, 4; Pl. X, 4; Fig. 31, p. 94)........................*Carpìnus** 94

156. Buds not angled; lf. scars little raised; bundle scars 3 in 3 groups; fr. a drupe (Pl. XVI, 4)...*Zélkova* 114

153. Phyllotaxy usually ½ on side (*plagiotropous*) brs., ⅖ on erect (*orthotropous*) shoots and long growths; pith rays fairly prominent in cross section with 10 × lens; pith small, often compressed, sometimes green; bud scales more or less striate; bundle scars 3; stipule scars narrow; in some spp. resinous warts on brts.; lenticels conspicuous, lengthened to transverse lines on older brts. (Pl. VIII and Pl. IX, 1, 2, 3)...........................*Bétula* 87

152. Stipule scars absent; shrubs or small trees.

157. Pith light brown; buds small, conical; lf. scars reddish; bundle scars crescent-shaped, compound; pith spongy in *S. paniculata*, usually chambered in *S. tinctoria* (Fig. 103a, p. 214)............................*Sýmplocos* 213

157. Pith white.

158. Buds very small, globular to cylindrical, pointing at wide angle from brts.; brts. slender, wand-like, often dying back a little; lf. scars minute; fr. often persistent into winter (*S. latifolia* and *tomentosa*)..........*Spiràea* 133

158. Buds larger, triangular, not pointing at wide angle from brt.; lf. scars much raised, hiding base of triangular, blunt-pointed bud (Fig. 58, p. 138)
Photinia villosa 139

---

* *Ulmus* and *Carpinus* are included here again because in young trees or vegetative shoots there are no fl. buds.

DISTINGUISHING CHARACTERS OF SPECIES

I. GYMNOSPERMAE — GYMNOSPERMS

Seeds naked, i.e., not enclosed in an ovary (Figs. 4, 6, 10, 15, 18 and Pls. I and II); trees or shrubs; lvs. needle- or scale-like (in *Ginkgo* fan-shaped), mostly evergreen; in *Ginkgo, Larix, Pseudolarix,* and *Taxodium,* lvs. are deciduous.   Includes Ginkgo, Yew, and Conifers (Pine Family).

GINKGOÀCEAE — GINKGO FAMILY

GÍNKGO

(Pronounced with either a soft or hard initial "G")

G. bilòba L. Ginkgo.   The only extant sp. of the genus.   Dioecious; lvs. fan-shaped, often two-lobed (*bi-lòba*), deciduous; short, thick, spur-like brts. along the main brs. (Fig. 3); seed with a thick, fleshy, outer

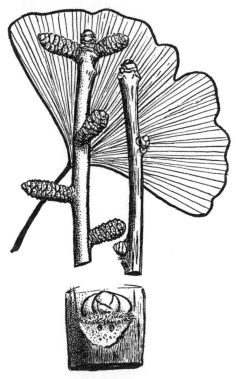

FIG. 3.   *Ginkgo.* Lf. and brts., nat. size; rt., brt. 1 yr. old; left, brt. several yrs. old, showing characteristic spur-like brts.   Apparently tip of older brt. is becoming a region of limited growth, as in the spur-like growths.   Below, bud and lf. scar × 4.

coat, when ripe superficially resembling a large yellow cherry. Native of China. Commonly cult. Remarkably tolerant of city conditions and therefore much planted as a street tree. No individuals now known in the wild state, and so sometimes called the "living fossil." Striking evidence of the antiquity of the sp. is furnished by the swimming sperms, reminiscent of the fern group of plants.

TAXÀCEAE — YEW FAMILY

Táxus — Yew

Seed solitary, surrounded by a red, fleshy aril. Lvs. evergreen, linear, green or yellow-green beneath and paler than above, but without white bands; without resin ducts; *distinctly petioled;* midrib of lf. slightly raised along upper surface. Dioecious, rarely monoecious; small trees, many vars. shrubby. Much cult.

The Yew is often confused with the Hemlock and Fir, but in the latter two genera the underside of the lf. has 2 white stripes, while in the Yew the same areas are green or yellow-green. However, as in the Hemlock, the lvs. have a distinct petiole which joins on to a little shoulder on the stem. See Fig. 4 and compare Fig. 5, c. But the most distinct character is the fruit, which is very ornamental, a red or deep pink, consisting of a fleshy growth, the aril, which in cuplike fashion surrounds the single seed (Fig. 4: 4 and 5). These "berries" when ripe are edible, but the seeds are deadly poison.

1. **T. canadénsis** Marsh. Canada Yew, also known as American Yew or Ground-hemlock, is a low straggling shrub in rocky woods, often under other evergreens. Readily distinguished from the Hemlock by the entirely green under-surface of the lf. Never very common but may be locally abundant. Nfd. to W.Va. and w.

Two spp. of Yew commonly cult. are Japanese Yew, *T. cuspidàta* Sieb. & Zucc. and English Yew, *T. baccàta* L. These two are very similar and are difficult to tell apart. The most reliable distinction is the *shape* of the lf. In the Japanese Y. the lf. comes to a sudden sharp point, or rather, narrows abruptly at the tip and is then produced into a point of appreciable length (Fig. 4: 1, 2, and 3). In the English Y. however, the lvs. gradually *narrow* to a point (Fig. 4: 7 and 8). Another difference easily seen is that the lvs. of the English Y. are more or less *shining* while those of the Japanese are dull. The Japanese is the more commonly cult. sp. Numerous vars. of these two spp. are cult.

Taxus mèdia Rehd., Anglojap Yew, is a hybrid of *T. cuspidàta* and *T. baccàta,* with its characters favoring the Japanese parent. Var. *hícksi,* columnar in habit with more or less erect brs. and radially spreading lvs., is often used for ornamental plantings and for hedges. Var. *hatfieldi* is also desirable for this purpose.

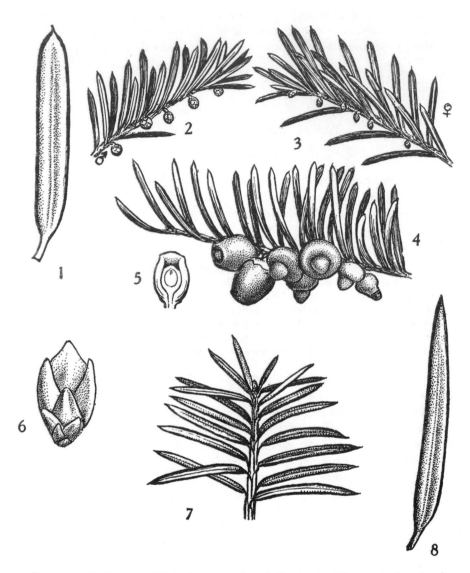

FIG. 4.   1–5; *Taxus cuspidata*, Japanese Yew. 1. lf. × 4; 2, ♂ brt., nat. size, showing staminate buds; 3, ♀ brt., nat. size, showing ovule-bearing fls. 4, ♀ brt. with various stages of development of fleshy aril around seed. 5, lengthwise section of "berry" showing position of seed in aril.   6–8, *T. baccata*, English Yew.   6, bud × 10, showing obtuse, scarcely keeled scales; 7, brt. nat. size.   8, lf. × about 3.

PINÀCEAE — PINE FAMILY

Trees or shrubs containing resin; lvs. mostly needle-like, linear, or scale-like (*Nadelbäume* of German foresters), evergreen[1] (*Pseudolarix, Larix,* and *Taxodium* have deciduous lvs.); monoecious (dioecious in Juniperus); fruit a *cone* with seed-bearing (*ovuliferous*) scales (Fig. 15, e and Pls. I and II), hence the term "cònifer," the cone-scales in Juniperus becoming fleshy and welded together to form a berry-like fruit (Figs. 22 and 23).

As represented in our area, the family may be divided into 3 tribes as follows:[2]

1. The Fir Tribe (*Abietíneae*) including *Abies, Pseudotsuga, Tsuga, Picea, Pseudolarix, Larix, Cedrus* and *Pinus*.

2. The Bald Cypress Tribe (*Taxodiíneae*) comprising *Sciadopitys, Sequoia, Sequoiadendron, Taxodium* and *Cryptomeria*.

3. The Cypress Tribe (*Cupressíneae*), comprising *Thuja, Cupressus, Chamaecyparis* and *Juniperus*.

ABIETÍNEAE — FIR TRIBE

Lvs. spirally arranged, fascicled, or whorled; needle-like or linear; cones constructed in general like those of *Abies*, the cone scales being spirally arranged about a central axis, each scale being borne in the axil of a bract. (Cf. Figs. 6, 10, 15).

ÁBIES — FIR

Lvs. linear, flat, with two white lines beneath; leaving a circular scar with a dot in the center when they fall (Fig. 5, a). Cones *erect*, falling off scale by scale, leaving the persistent axis.

1. **A. balsámea** (L.) Mill. Balsam Fir. The native fir of northern New England and northern N.Y., also along the mountains to W.Va. and w. Does not take kindly to cult. where hot summers are the rule. May be recognized by its very resinous and fragrant buds and fairly short lvs. (¾–1 in. long), rounded, or bluntly short-pointed, sometimes notched, at tip, disposed usually (but not always) in fairly flat series on each side of the br. (though in reality inserted spirally). Very popular

---

[1] The members of the Pine and Yew families are sometimes erroneously called "The Evergreens." But there are many broad-leaved trees and shrubs which are also evergreen, e. g., many members of the Heath Family, such as *Rhododendron*, Mountain-Laurel, *Pieris*, and others; and the Hollies.

[2] Some authorities have raised these tribes to the rank of families, thus:

1. *Pinaceae*, Pine Family (= *Abietineae*)
2. *Taxodiaceae*, Bald Cypress Family (= *Taxodiineae*)
3. *Cupressaceae*, Cypress Family (= *Cupressineae*)

as a Christmas tree, but the Douglas-Fir (*Pseudotsuga*) is said to hold its lvs. longer.

2. *A. cóncolor (Gord. and Glend.) Lindl. White Fir. Occasionally cult.; while it also has resinous buds it is characterized especially by its *long* (1½–2½ in.) *curved, bluish* lvs. Colo. to Calif. and n. into Ore. also s. through N.Mex. and Ariz. into northern Mexico and Lower Calif. The only Fir in the dry areas of the Great Basin, southern N.Mex. and Ariz.

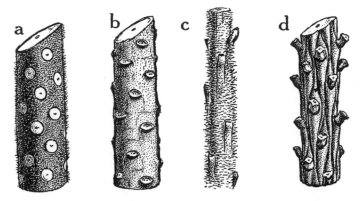

Fig. 5. Characteristic appearance of leafless twigs of 4 genera of the Fir tribe. a, *Abies;* b, *Pseudotsuga;* c, *Tsuga;* d, *Picea,* all × 4. Note spiral arrangement (see p. 5).

and of the mountain forests of southern Calif. Hence it endures well dry, hot, summer conditions in cities.

3. A. nordmanniàna Spach. Nordmann Fir. Lvs. notched at tip; buds *not resinous.* Native in Greece, Asia Minor and the Caucasus region. The most commonly cult. sp. Grows well in our region, even in smoky cities.

PSEUDOTSÙGA — DOUGLAS-FIR

*P. menziesi (Mirb.) Franco (P. taxifòlia (Poir.) Britton). Douglas-fir. Red-fir. Buds red-brown, smooth, rather large, *sharp-pointed;* lvs. linear, with two grayish lines beneath; lf. scars circular, somewhat as in *Abies*, but *tilted at an angle* from the twig (Fig. 5, b). The characters of the buds and lf. scars are very distinct and are important for the recognition of this sp. Native in the mountains of w. N.A. Much cult. in several vars. and grows well here. In this genus the bracts, which in the fir tribe always subtend the cone scales, grow out beyond the latter. Here the bracts are 3-pointed (Fig. 6, a).

Tsùga — Hemlock

Lvs. linear, flat, with two white lines beneath, and *distinct* short pet-ioles (Figs. 7 and 8). Lvs. borne on slight elevations on stem, which are not as pronounced as in *Picea* (Fig. 5, c). The lvs. of *Abies*, the fir, which resembles this, have no distinct petioles, although they are much contracted at the base.

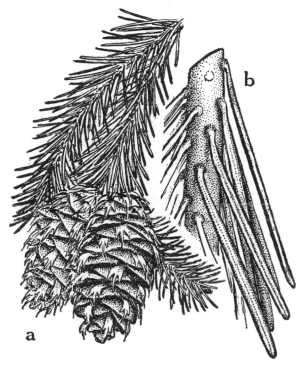

Fig. 6. *Pseudotsuga taxifolia*, Douglas-Fir. a, brt. with cones, nat. size; b, part of brt. with lvs. × 4.

1. **T. canadénsis** (L.) Carr. Eastern Hemlock. Lvs. finely toothed on the margin (seen with a lens), blunt at tip; usually several small lvs. occur upside down along brt. (Fig. 8). Brts. pubescent. The only common native sp. in the eastern U.S.; grows on rocky ridges and in rocky ravines. One of our most beautiful conifers. Weeping and many other vars. occur in cult. See list in S.P.N., pp. 629, 630.

2. The Carolina H., *T. caroliniàna* Engelm., is very similar, but the lvs., which are not toothed, are disposed at various angles on the twigs giving the branches as a whole a rougher appearance. The cones are slightly larger. A rare tree in the wild. Se. Va. to n. Ga.; sometimes cult. in the north.

PÌCEA — SPRUCE

Lvs. in following spp. (except in *P. omorika*) 4-sided, acicular (Fig. 9) borne on distinct *short peg-like projections on the stem*, which remain when the lvs. are shed, giving brt. a rough appearance and feel (Figs. 5, d, and 10, 7); *cones pendulous*, their scales persistent (Pl. I).

1. The Tigertail S., *P. polìta* (Sieb. & Zucc.) Carr. has dark-brown, pointed buds, the scales of which long persist as a blackish sheath at the base of brts.; often curved, shiny, stiff, spiny-pointed lvs. ¾–1 in. long.   Brts. stout, pale yellow, glabrous. Occasionally cult. Japan.

 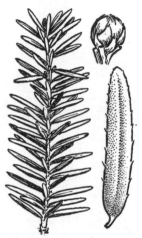 

FIG. 7. *Tsuga canadensis*, Eastern Hemlock. Brt. with cone nat. size.

FIG. 8. *Tsuga canadensis*, left, brt. nat. size showing characteristic upsidedown lvs.; rt., above, bud × 4; below, lf. showing serrulate margin × 4.

FIG. 9. *Picea.* Part of lf. with section showing 4 sides, × 4.

2. **P. ábies** (L.) Karst.   (*P. excélsa* Link) Norway S.   The commonly cult. spruce, recognized by its long cones (4–6 in.) (Pl. I, 5) and pendulous brts., which are glabrous.   Has a multitude of vars. and forms.   Native in northern central Eu.

3. **P. orientàlis** (L.) Link.   Oriental S.   Lvs. very short (¼–⅜ in.), blunt, appearing somewhat as if bevelled like a chisel, shining; brts. pale brown, short-pubescent; cones 2½–3½ in. long.   Native in Asia Minor and the Caucasus region.   Occasionally cult.

4. **P. rùbens** Sarg.   Red S.   In this and the next sp. the terminal bud has a ring of awl-shaped scales at its base and the brts. are pubescent; cones oval, 1–2½ in. long, purplish or light green when young, red-brown when mature, *deciduous;* cone scales entire, or only slightly toothed. (Pl. I, 2 and Fig. 10, 3).   P.E.I. and N.S. to mts. of N.C. and Tenn.

5. **P. mariàna** (Mill.) B.S.P. Black S. Terminal buds with a ring of awl-shaped scales at base, and brts. pubescent. Cones smaller than in Red S., ½–1½ in. long, and egg-shaped or nearly spherical, gray-brown, *persistent* — sometimes remaining on the brs. 20 or 30 yrs. Cone scales finely toothed. (Pl. I, 1 and Fig. 10, 1 and 2). Prefers cool temperatures at high elevations, or northern latitudes; when growing in lowlands usually occurs in cold swamps. Lab. to Nfd. to Alaska and along Appalachians to n. W.Va. and w. Takes unkindly to cult.

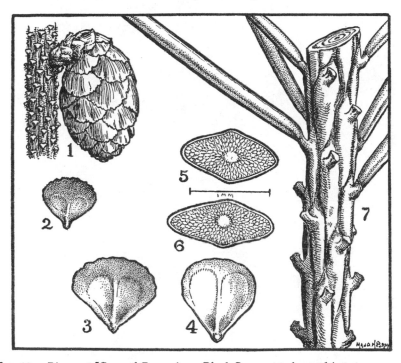

FIG. 10. *Picea.* 1, Cone of *P. mariana*, Black Spruce × about 1½; 2–4, cone scales × about 1½; 2, Black S.; 3, *P. rubens*, Red S.; 4, *P. glauca*, White S.; 5 and 6, cross sections of lvs. of *P. abies*, Norway S. and *P. omorika*, Serbian S. respectively, enlarged; 7, brt. of Norway S. enlarged. Courtesy of National Audubon Soc. and Bklyn. Botanic Garden.

6. **P. glauca** (Moench) Voss. White S. Brts. *glabrous*, gray or pale brown. Lvs. blue-green and more or less glaucous, with disagreeable odor when bruised, the strongest-smelling individuals known locally as "cat" or "skunk" spruce; cones *cylindrical*, longer than in Red or Black S. (1½–2 in.), before maturity bright green, tinged with red, finally becoming glossy light-brown, the scales *entire*. (Pl. I, 3 and Fig. 10, 4). Northern N.E. and n. N.Y. and w.: Lab. to Alaska. Several cult. vars.

PLATE I. Cones of Spruces.

1. *Picea mariana*, Black S. 2. *P. rubens*, Red S. 3. *P. glauca*, White S. 4. *P. pungens*, Colorado S. 5. *P. abies*, Norway S. Cones of Red and of White S. furnished by Dept. of Forestry, Univ. of Maine. All about ½ nat. size. Photo by L. Buhle, 2-23-43. Courtesy Brooklyn Botanic Garden and National Audubon Society.

The Red, Black, and White S. are the only native spruces in this area. A fourth sp., the Norway S., is common, but always cult. and readily distinguished by its long cones (4–6 in.), usually near the top of the tree, and its multitude of short pendulous branches, which, together with its somber, very dark green, almost black color, give to the whole tree a funereal aspect.

The Red and the White S. are often found together in n. N.E. and N.Y. and then the blue-green cast of the White S. foliage contrasts markedly with the dark green of the Red. Moreover, the White S. has glabrous brts., while those of the Red are pubescent. The cones, too, are quite different; those of the White S. being cylindrical with entire or nearly entire cone scales, while in the Red S. the cones are smaller, more ovate ("elongated egg-shaped"), and the cone scales somewhat jagged or irregularly toothed.

In both the Red and the White S. the cones fall off the tree early (deciduous) while in the Black S., which also may be encountered in the same general regions, the small, almost spherical cones *persist* on the tree for many years. The Black S. has pubescent brts., and in the more southern parts of its range is often a swamp sp.

The ranges of these 3 native eastern spruces are interesting. In the extreme northeast the Red and the White are commonly found together, while in the U.S., at least, the Black is not abundant. But from the common meeting place of all 3 in the northeast the Black and the White follow a broad trans-continental belt extending northwestward even into Alaska, while the Red, strictly an eastern sp., leaves them and follows the Appalachians as far south as N.C. and Tenn. However, the Black does occur sparingly as far south as W.Va. For maps showing this distribution see Collingwood, G. H. "Knowing your trees" (10, pp. 48–53).

7. *P. púngens Engelm. (*P. parryàna* Sarg.) Blue or Colorado Spruce. Lvs. stiff, sharp, blue; brts. glabrous, cones 2½–4 in. long. (Pl. I, 4). Often cult., with many vars., of which the Koster S. (var. *kôsteri*) in which the lvs. are very light colored, is a popular one. Native in mts. of Colo. to N.Mex., Utah and Wyo. mostly below Engelmann Spruce belt. Probably the best sp. for planting in dry climates.

8. The Engelmann S., *P. engelmánni* Parry, is similar to *P. pungens* and sometimes also cult. in the e.U.S. Also has bluish lvs. but can be distinguished by *pubescent* brts., softer, more flexible lvs., and deep blue-green color.

9. The Serbian S., *P. omórika* (Pancic) Purkyne, has flattened lvs. (Fig. 10, 6) and white bands on *upper* surface of lf. with distinct green midrib; brown, pubescent brts., and winter buds with awl-shaped scales at base. Native in s.e. Eu. Sometimes cult. Hardy but slow growing.

10. The Koyama S., *P. koyamài* Shiras., resembles Norway S. in general habit, but brts. are thicker and reddish; lvs. slightly flattened with 2 white bands above; cones cylindrical, 1½–4 in. long. Japan, Korea. Fine mature specimens at Arnold Arboretum.

PSEUDOLÁRIX — GOLDEN-LARCH

Lvs. deciduous, borne mainly in whorl-like clusters at the ends of spur-like brs. Much like the true larch (*Larix*), but with *cone scales deciduous*, leaving the axis of the cone standing on the tree: the spur-like brs. and the lvs. are usually longer than in *Larix* (Fig. 11, b). One sp.

P. amàbilis (Nels.) Rehd. Golden-larch. A beautiful tree, easily cult., lvs. turning bright yellow in fall. China.

LÁRIX — LARCH

Lvs. deciduous, borne mainly in whorl-like clusters at the ends of short, spur-like brs. (Fig. 11, a), *cones persistent.*

1. L. decídua Mill. (*L. europaea* DC.) European L. Cones ¾–1½ in. long; cone scales slightly downy outside, 40–50 to a cone. The sp. commonly seen in parks and private grounds.

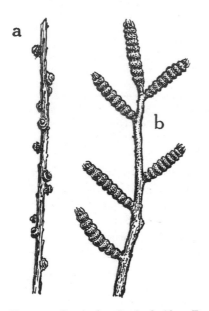

FIG. 11. Brts. of a, *Larix decidua,* Eu. Larch, 2 yrs. old, and b, *Pseudolarix amabilis,* Golden-Larch, 9 yrs. old, showing characteristic short brts. or growths.

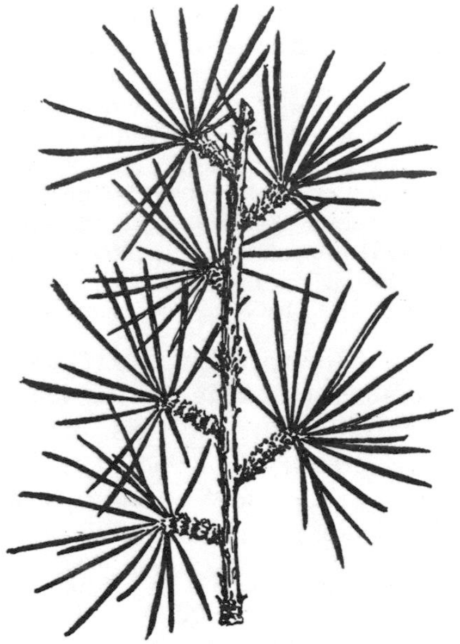

FIG. 12. *Cedrus atlantica,* Atlas Cedar. Brt. 5 yrs. old, showing characteristic short growths.

2. **L. larícina** (DuRoi) K. Koch. Eastern L., Tamarack, Hackmatack. Cones ½–¾ in. long; cone scales smooth outside, 12–15 to a cone; lvs. about as in last (¾–1½ in.). Prefers moist soil and grows in abundance in the swamps of the northern States. Occasionally takes kindly to cult. in drier soil. Lab. and N.E. s. to W.Va. and nw. to Alaska.

3. **L. leptólepis** (Sieb. & Zucc.) Gord. Japanese L. Differs from above 2 spp. in having white bands on the *under* sides of the lvs. (easily seen with a hand lens) and in the cone scales being *recurved.* Japan. Occasionally cult. Handsome and grows rapidly.

Cédrus — Cedar[1]

Lvs. acicular, 3-sided, mainly in dense whorl-like clusters on short, spur-like brs. (Fig. 12). Much like *Larix*, but lvs. are *evergreen* and cones much larger, erect on the brs. (Fig. 13).

Four spp., 3 given below, the 4th, *C. brevifòlia*, a native of Cyprus, with very short lvs. (about ¼ in. long) and small cones. It is interesting to note that all these spp. are not far from latitude 30 N. proceeding from w. in a rather straight line to e., beginning in N. Africa, and ending in the Himalayas. The spp. are so similar that they have been considered as races of one sp.

Fig. 13. Brts. and cone of *Cedrus atlantica* nat. size.

1. C. atlántica Manetti. Atlas Cedar. Habit of tree is stiff; lvs. bluish-green, usually less than 1 in. long; brts. *short pubescent;* cones 2–3 in. long, about 1½ in. in diam. (Fig. 13). Blooms *Sept. to Oct.* Native

[1] The term "cedar" may confuse the beginner. There are so many cedars: which is the real cedar? If one is to judge from the botanical name, *Cedrus* is the true cedar. *Cedrus líbani*, the Cedar-of-Lebanon, was known far back in Biblical times and was used in the construction of Solomon's temple. But cedar, *per se*, is "any one of a large number of trees having fragrant wood of remarkable durability" (Webster's Dictionary). Many of these trees belong to different genera in the Pine Family. Thus we have *Chamaecyparis*, yielding White-cedar, and *Thuja*, also known as White-cedar. But SPN has allotted the name "Arbor-Vitae" to *Thuja*, thus clearing up the situation somewhat. *Juniperus* gives the wood known as Red-cedar (pp. 71, 72). As always, the botanical names are the important ones, since they connote certain botanical characters which the common name "cedar" does not.

in N. Africa. Named from Atlas Mts. Hardy in extreme s. N.E., s. N.Y. & N.J. Rather rare in cult. but becoming popular, especially in the var. *glauca*.

2. C. líbani Loud. Cedar-of-Lebanon. Habit more graceful and spreading; lvs. dark or bright green, a little longer than in last; brts. glabrous or slightly pubescent; cones 3–4 in. long, 1½–2½ in. in diam. Asia Minor. Some strains hardy in s. N.E.; fine specimens have been grown in the vicinity of N.Y. City. These last two spp. much cult. in England.

The Deodar Cedar, *C. deodàra* (Roxb.) Loud., is a beautiful tree with longer lvs. (2 in.), drooping and densely pubescent brs. Not hardy in our area but cult. in Southern States; e.g., in Charleston, S.C., Savannah, Ga. Himalayas.

Pìnus — Pine

Lvs.[1] long, needle-like, evergreen, in our spp. in *fascicles* of two or more.

Lvs. in fascicles of 5, not sheathed at base.
  Brts. glabrous or nearly so.
    Brts. glaucous; lvs. long (4–7 in.), cult.........................3. *P. griffithi* 64
    Brts. not glaucous; lvs. shorter (2½–5 in.), native (Fig. 14)......1. *P. strobus* 64
  Brts. puberulous to tomentose; cult.
    Brts. thick and tomentose.....................................2. *P. cembra* 64
    Brts. more slender, puberulous or nearly glabrous, lvs. in tufts at ends of brts.
                                   4. *P. parviflora* 64
Lvs. in fascicles of 3, sheathed at base.
  Lvs. 6–9 in. long. Native s. N.J. to Fla. and w...................7. *P. taeda* 65
  Lvs. 3–5 in. long. Native N.B. to Ga. and w....................6. *P. rigida* 65
Lvs. in fascicles of 2, sometimes in 3's in *P. echinata*.
  Shrubby, lvs. short (1¼–3 in.). Much cult....................14. *P. mugo* 66
  Trees.
    Lvs. with blue-green cast, bark orange in upper part of tree; cult. and naturalized..11. *P. sylvestris* 66
    Lvs. green.
      Young brts. glaucous, at least when young.
        Buds not, or little, resinous; lvs. 3–5 in. long.............8. *P. echinata* 65
        Buds resinous; lvs. 1½–3 in. long...................13. *P. virginiana* 66
      Young brts. not glaucous.   (See also *P. densiflora*) 64
        Buds brown, resinous.
          Lvs. very short (½–1½ in.).....................12. *P. banksiana* 66
          Lvs. longer.
             Lvs. thickish, dull (3–6 in.), cult......................9. *P. nigra* 65
             Lvs. slender, shining (4–6 in.), native................5. *P. resinosa* 64
        Buds white, cylindrical, not resinous; lvs. 3–5 in., cult....10. *P. thunbergi* 65

---

[1] Strictly, these are the *secondary* lvs. borne in a whorl (fascicle) on a very short br., which is subtended by a scale-like *primary* lf.; these scale lvs. are more prominent in the bud stage, there functioning as bud scales. In the one-yr.-old seedling the primary lvs. are the foliage lvs., the secondary lvs. usually appearing only when the seedling is 2 or 3 yrs. old. Then the primary lvs. gradually assume the form of scales.

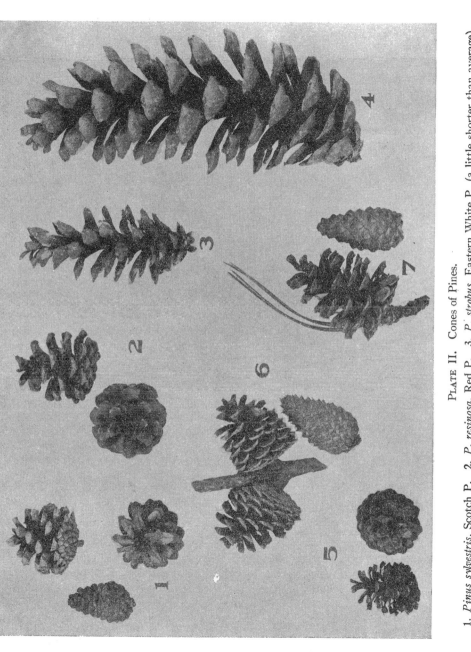

PLATE II. Cones of Pines.

1. *Pinus sylvestris*, Scotch P. 2. *P. resinosa*, Red P. 3. *P. strobus*, Eastern White P. (a little shorter than average).
4. *P. griffithi*, Himalayan P. 5. *P. mugo*, Swiss Mountain P. 6. *P. rigida*, Pitch P. 7. *P. nigra*, Austrian P. All about
½ nat. size. Photo by L. Buhle. Courtesy Brooklyn Botanic Garden.

1. **P. stròbus** L.   Eastern White P.   The common 5 lvd. pine native in this area (Fig. 14).   Lvs. 2½–5 in., usually about 4 in. long, not or only slightly drooping, *serrulate*.   Cones long (3½–8 in.) usually 4–6 in., chiefly near top of tree.   (Pl. II, 3).

The 5-needled pines form a special group of spp. within the genus, characterized by soft wood as well as by having the lvs. in fascicles of 5, hence called the Soft P. The following 5-needled pines are more or less common in cult. in this area: 2. *P. cembra* L. Swiss Stone P., with rather thick, brownish, tomentose brts., lvs. serrulate, 2–5 in. long and wingless seeds.   Alps, Russia and Siberia: 3. *P. gríffithi* McClelland, Himalayan P. (*P. excélsa* Wall. and *P. nepalénsis* De Chambray) easily recognized by its

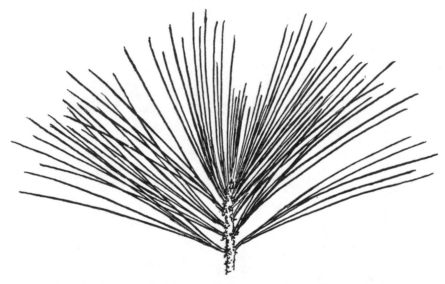

Fig. 14.   *Pinus strobus*, Eastern White Pine.   Brt. and lvs. nat. size.

long (4–8 in.) *drooping* needles, a magnificent tree of the Himalayas, w. to Afghanistan; sometimes suffers in severe winters (Pl. II, 4): 4. *P. parviflòra* Sieb and Zucc. Japanese White P., usually, in cult., a small tree with short lvs., the brs. symmetrically arranged in tiers, and lvs. in tufts at the ends of brts.   These last two spp., together with *P. strobus*, have *winged* seeds and *serrulate* lvs.   *P. fléxilis* James, Limber P., a 5-needled pine, of the w. and sw.U.S., rarely cult., has entire lvs. and nearly or quite wingless seeds.

5. **P. resinòsa** Ait.   Red P., also known as Norway P.   Lvs. in 2's, long (4–7 in.), closely resembling the commonly planted Austrian P. (No. 9, p. 65) but lvs. more slender, and shining, buds resinous.   Bark has a reddish tinge, hence the common name.   Cone scales without the short prickle characteristic of the Austrian P.   Nfd. to Man., s. along mts. to W.Va., and w.   (Pl. II, 2).

P. densiflòra Sieb and Zucc., the Japanese Red P., is similar, but buds are not resinous and young brts. are glaucous.   Rarely cult.

P. pungens, the Table-Mountain P., with stout, short lvs. in 2's or 3's and persistent cones with strong, hooked spines, occurs naturally in uplands from s. N.J., s. and w.

6. **P. rígida** Mill. Pitch P. Lvs. in 3's, cones with sharp prickles, on any part of the tree, and persistent. One of our commonest pines; grows in poor, often sandy soil, and useful for planting on such sites; with a scraggly but picturesque habit. Me. to Ga. (Pl. II, 6).

7. The Loblolly P., **P. taèda** L., with slender lvs. in 3's (rarely in 2's) (5–10 in.), long basal sheaths, occurs from s. N.J. to Fla. and w.; also 8. **P. echinàta** Mill., Shortleaf P. with 2 or sometimes 3 lvs. in a fascicle (3–5 in.), a valuable lumber tree, yielding, along with **P. taèda,** *palústris* and others, the so-called "southern" or "S.Carolina" or "Southern Yellow Pine" lumber, occurs toward the south of our area, L.I. to Fla. and w.

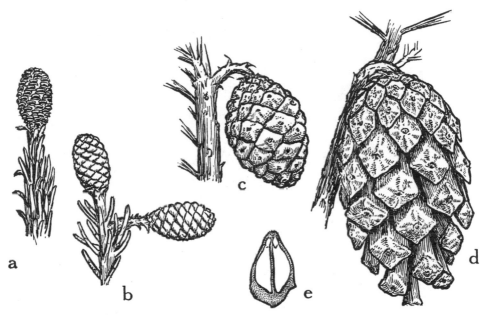

FIG. 15. *Pinus sylvestris*, Scotch Pine. a, ♀ fl. (ovulate cone) unfertilized; b, fertilized cones, end of 1st yr.; c and d, stages in development of cone; d, cone opening to discharge seeds, end of 2nd or beginning of 3rd yr.; e, seed scale showing 2 winged seeds.

9. P. nìgra Arnold. Austrian P. Lvs. in 2's, long (3–6 in.), dark green. Unlike the Red P. which it closely resembles, the cone scales usually have a short prickle. Native in s. Eu. One of most commonly cult. pines; endures city conditions well (Pl. II, 7). Several vars.

A sp. closely resembling the Austrian P. is (10) the Japanese Black P., *P. thúnbergi*, with very dark green, stiffer lvs. (3–5 in. long) *white, cylindrical* buds, the bud scales deeply fringed. Sometimes cult. Best conifer for coastal reforestation. High saline tolerance and resistance to insect pests. See Littlefield, E. W. *Jour. Forestry* 40: 566–573. 1942, and Jones, Bassett. *Natl. Hort. Mag.* 9: 181–190. 1930.

11. P. sylvéstris L.　Scotch P.　Lvs. in 2's, short or of medium length (1½–3 in.), blue-green; young bark with an orange color which usually shows on upper part of trunk.　Native in Eu. and Siberia.　One of most commonly and easily cult. pines and often naturalized (Pl. II, 1 and Fig. 15, p. 65).

12. P. banksiàna Lamb.　Jack P.　Lvs. in 2's, very short and thick (1–1½ in.), cones 1–2 in. long, with very short prickles or none, lopsided and *usually curved toward the brt.* on which they are borne, persistent for many years.　Sometimes a tree (60 ft.) but more often smaller or shrubby.

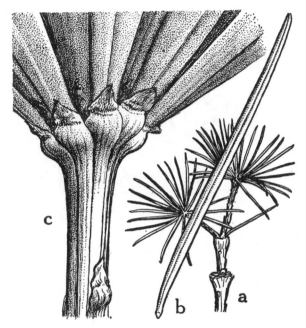

Fig. 16.　*Sciadopitys verticillata*, Umbrella-Pine.　a, brt. and lvs. × about ⅕; b, lf. nat. size, c, part of brt. × 5, showing scale lvs. (primary lvs.) above, in apparent whorl with secondary lvs. in their axils; below, single scale lf. of spiral arrangement.

N.S., c. Quebec, n. N.E. & N.Y. and w.　Comparatively rare in our region; abundant in Mich. and n.w. in Canada, covering large tracts of sandy, sterile soil.

13. P. virginiàna Mill. Virginia P., Jersey or Scrub P.　Lvs. in 2's, of medium length (1½–3 in.), brts. covered with *glaucous bloom*, buds very resinous, cones 2–3 in. long, with *persistent prickles*.　A small tree (30–40 ft.) L.I. & S.I., N.Y. and N.J. to c. and w. Pa., s. to Ga. and w.

14. P. mùgo Turra.　Swiss Mountain P. (Pl. II, 5).　Lvs. in 2's, stout and short (1¼–3 in.).　Usually a low prostrate shrub, in age larger but still shrubby. Often used in landscape gardening and easily recognized by its short, thick, dark green lvs. in 2's as well as by its prostrate habit. However, vars. and forms with a more erect habit occur (Pl. II, 5).

TAXODIÍNEAE — BALD-CYPRESS TRIBE

Lvs. linear; spirally arranged, whorled apparently in part in *Sciadopitys*, and distichous on short brts. of *Taxodium;* cone-scales without distinct bracts.

SCIADÓPITYS — UMBRELLA-PINE

S. verticillàta Sieb. & Zucc. Umbrella-pine. Lvs. (said to be the morphological equivalent of 2 connate pine needles) long, linear, evergreen, borne in whorls, each lf. in the axil of a little scale lf., corresponding, respectively, to the secondary and primary lvs. in *Pinus* (Fig. 16). Native in Japan. Often cult. Slow-growing, handsome, pyramidal trees.

The Redwood, *Sequoia sempervìrens,* and the Giant Sequoia, *Sequoia gigantea,* are two famous trees of the Pacific Coast region which belong to the *Taxodium* group. *Mètasequòia,* a recently discovered ancient tree in sw. China, is closely related to these 2 spp. (P. & G. 4: 231–235. 1948).

FIG. 17. *Taxodium distichum,* Bald-cypress. a, brt. nat. size, showing radially arranged, persistent lvs. and 3 deciduous short growths or brts. with distichous lvs. b, brt. × 5, showing radially arranged lvs.

TAXÒDIUM — BALD-CYPRESS

Brts. of two kinds: those near tip of shoot persistent and with buds in the axils of the lvs.; those on lower part of shoot without axillary buds and deciduous.[1] Buds very small, globular, scaly. Lvs. small, flat, yellow-green on both sides; those of persistent brts. projecting radially; those on deciduous brts. 2-ranked (Fig. 17, a and b).

T. dístichum (L.) Richards. Bald-cypress. Twigs more or less horizontal. Common in swamps of se. U.S. and low river bottom-lands. Grows well in moist soil. S. Del. to Fla. and w. Also cult. n. (e.g. Sharon Springs, N.Y.).

Some good specimens of the Pond Bald-cypress, * T. ascéndens Brongn., Va. to Fla. and w., may be seen at the N.Y. Botanical Garden, to the southeastward of the conservatories where they may be compared with T. distichum near by. The Pond Bald-cypress has more upright brs. and narrower lvs. than the common sp.

Fig. 18. *Cryptomeria japonica.* Brt. and cone nat. size.

CRYPTOMÉRIA

C. japónica D. Don. Cryptomeria. Lvs. very short, awl-shaped, curved, evergreen, strongly decurrent on the stem (Fig. 18). A timber as well as an ornamental tree in its native Japan. Several cult. vars.

CUPRESSÍNEAE — CYPRESS TRIBE

Lvs. usually small and scale-like, sometimes acicular;[2] decussately opposite or in whorls of 3; cone scales opposite or in whorls of 3.

THÙJA — ARBOR-VITAE

Lvs. tiny and scale-like; those on sides of brt. keeled; those on upper and lower surfaces flat. Cones small, constructed on the plan of the pine or spruce cone.

1. T. occidentàlis L. Eastern Arbor-Vitae, Northern White-cedar. Cones to ½ in. long; each flat lf. with a resin gland (Fig. 19). Twigs have characteristic resinous taste. N.S. to Man. and s. along mts. to N.C. and w. Commonly cult. in many vars.

2. T. orientàlis L. Oriental Arbor-Vitae. Easily recognized by the vertical planes in which the brts. are disposed, and by the peculiar, rather large hook on

[1] The deciduous brts. correspond morphologically to the lf. fascicles (short brts.) in PINUS and to the short brts. in LARIX and CEDRUS.

[2] In this tribe the juvenile lvs., that is, those of young plants, are awl-shaped or needle-like (cf. Figs 21, 3, and 22) pointing outward from the stem, and so young seedlings present an entirely different aspect from older trees. In some horticultural forms

the cone-scales.   Cones ovoid, ½–1 in. long (Fig. 20).   N.W. China, Korea.   Commonly cult. in many vars.

The Western Red-cedar, *Thuja plicàta*, Donn., a large timber tree, to 180, rarely 200 ft., native from Alaska to N.Calif. and Mont. with inconspicuously glandular lvs., and brts. whitened below, and the Japanese A., *T. stándishi*, (Gord.) Carr., with glandless lvs., rather thick, scarcely flattened brts., whitened beneath, are sometimes cult. in our area.

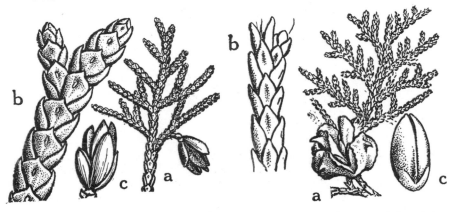

FIG. 19.   *Thuja occidentalis*, Eastern Arbor Vitae.   a, brt. with cone nat. size; b, brt. showing 2 forms of lvs., the flat lvs. with resin glands × 5; c, cone × 2.

FIG. 20.   *Thuja orientalis*, Oriental Arbor Vitae.   a, brt. showing lvs. and hooked cone nat. size; b, brt. × 5; c, seed (no wings) × 4.

CHAMAECÝPARIS — WHITE-CEDAR[1]

Lvs. small and scale-like (awl-shaped or linear in some vars.); in the Japanese spp. listed below, whitened beneath.   Lvs. and brs. much resemble those of *Thuja* but the cones are quite different.   Cones composed of shield-shaped scales (the stalk of the scale being produced from the center of its under surface), the scales being fitted together in such a way that the whole forms a little ball (Fig. 21, 1, c.).   The Japanese spp. and vars. are much cult.

of *Thuja* and *Chamaecyparis* this juvenile form of lf. has been retained, usually as a result of horticultural treatment, and has become permanent in the mature tree; such forms are known in the trade as *Retinospora* (original spelling *Retinispora*).   For example, *Chamaecyparis pisifera* var. *plumosa* is sometimes called in nursery catalogs *Retinospora plumosa*.   However, botanists do not recognize the name *Retinospora*.

In *Juniperus virginiana* and related spp. both the mature scale-like and the juvenile awl-shaped lvs. will usually be found in different parts of the same tree, but in *J. communis* the juvenile form of lf. has persisted into the mature stage of the plant and is the only kind of lf. present.

[1] See note 1, p. 61.

FIG. 21. *Chamaecyparis*, 1, *C. thyoides*, Southern White-Cedar; a, brts. and cones nat. size; b, brt. × 4; c, cone × 4; 2, *C. pisifera*, Sawara Cypress, brt. nat. size with portion × 5; 3, *C. pisifera* var. plumosa, Plume Sawara C., brt. nat. size, with portion × 5.

1. **C. thyoìdes** (L.) B.S.P.   Southern White-cedar, Atlantic White-cedar.   Lvs. not whitened beneath as in exotic spp.   Cones bluish purple; about same size as in no. 2.   Does not take as kindly to cult. as Japanese spp.   Grows in swamps, not far from coast.   Me. to Fla., w. to Miss.   (Fig. 21, 1).

2. **C. pisìfera** (Sieb. & Zucc.) Endl.   Sawara Cypress.   Cones small (¼ in. in diam. or slightly larger), brown.   Brts. with white marks beneath.   Japan.   (Fig. 21, 2).

Vars. most commonly cult. (also known in the trade as "Retino-sporas") are:

C. pisìfera var. plumòsa Beissn.   Plume S., with feathery or plume-like brts. and awl-shaped lvs.   (Fig. 21, 3).

C. pisìfera var. squarròsa Beissn. & Hochst. Moss S., with spreading, linear lvs.

C. pisìfera var. filìfera Beissn.   Thread S., with gracefully drooping, threadlike brs.

C. obtùsa (Sieb. & Zucc.) Endl., Hinoki Cypress, with brts. in horizontal planes, *obtuse* lvs. with white marks below, is often cult. (Japan); also **C. lawsoniàna* (A. Murr.) Parl., Lawson Cypress, with acute or acutish, glandular lvs., white markings below often indistinct.   Ore. to Calif.

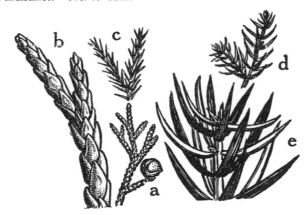

FIG. 22.   *Juniperus.*   a-c, *J. virginiana*, Eastern Red-cedar or Savin.   a, brt. with "berry" nat. size; b, × 5; c, juvenile growth nat. size; d-e, *J. communis*, Common Juniper, d, brt. nat. size; e, × 3.

Juníperus — Juniper

Cone berry-like, the scales having become fleshy and welded together.

1. **J. virginiàna** L.   Eastern Red-cedar or Savin.   Lvs. scale-like, in 4 rows lengthwise on the brts., making the brts. appear 4-sided; but on young trees or on vigorous shoots, lvs. are needle-like, sometimes in 3's

but often opposite. Fr. about ¼ in. in diam., bluish with a bloom (Fig. 22, a–c).   Can. to Fla. and w.

2. **J. commùnis** L.   Common Juniper.   Has *all its lvs. needle-like* and in whorls of 3 and often occurs as a shrub in circular patches (var. *depressa* Pursh) or sometimes tree-like (var. *erecta* Pursh).   Fr. berry-like, about ¼ in. in diam., bluish or black, slightly glaucous.   Can. s. to Md., mts. of Ga. and w. (Figs. 22, d, e and 23).   A cosmopolitan sp.; also in Eu. and Asia.   Cult. in many vars.

FIG. 23.   *Juniperus communis*, Common Juniper.   Brts. nat. size.

The Creeping J., **J. horizontàlis** Moench, which grows close to the ground with long trailing brs., occurs from Nfd. to sw. Vt., nw. N.Y. and w.

The Chinese J., *J. chinénsis* L., with lvs. often markedly of 2 kinds, the scale-like obtuse and the acicular ones (often in 3's) spiny pointed with 2 white bands *above*, with brown fr., is sometimes cult.; and its var. *pfitzeriàna* Spaeth, Pfitzer J., with wide-spreading, low growing brs. and a general rough, unkempt appearance, is especially popular for landscape effects.

II.   ANGIOSPERMAE — ANGIOSPERMS

Seeds enclosed in an ovary which, when ripe, becomes the fruit.   Lvs. mostly broad (rarely needle- or scale-like); deciduous (rarely evergreen). Here belong the so-called "broad-leaved trees and shrubs."

CLASS I.   MONOCOTYLEDONS

In this class woody plants are scarce, notable ones being the palms and palmettos, not hardy in our area.   Moreover, their wood is not formed by growth and division of a cambium ring in the way that obtains in the gymnosperms and dicotyledons, but usually by a thickening and hardening of tissues already present.   Represented here by one genus which may be called woody.

LILIÀCEAE — LILY FAMILY

SMÌLAX — GREENBRIER, CATBRIER

In woods and along fences and borders of fields two woody, climbing species of *Smilax* are common in this region. Being closely related to the lilies, they have no central pith, but the vascular bundles are distributed throughout the stem (Fig. 24). These two species have underground rootstocks, and aerial stems, the latter usually green and prickly, and have a pair of tendrils (metamorphosed stipules) near the base of the lf. stalk.

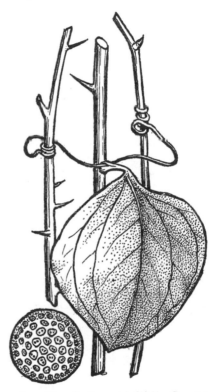

1. [00]**S. rotundifòlia** L. Common Greenbrier, Bullbrier. (Fig. 24). The commoner of the two; can be recognized by its thicker stems, stronger prickles, and *lvs.* rounded and *shining* on *both* surfaces. N.S. to Fla. and w.

2. [00]**S. glauca** Walt. Sawbrier (Cat Greenbrier SPN). Has more slender stems and prickles and ovate *lvs.* which are *glaucous* beneath. E. Mass. to Fla. and w.

Other woody spp., not so common in our area are: [00]**S. bòna-nóx** L., Saw Greenbrier, with angular, often square brts., lvs. (often fiddle-shaped) green and shining on both surfaces and black berries with a bloom. Nantucket, Mass. and Del. to Fla. & w. [00]**S. híspida** Muhl., Chinaroot, with blackish, bristly prickles and black berries, Conn. to Ala. and w., locally common in N.Y. State; and [00]**S. laurifòlia** L., Laurel-lvd. Greenbrier, with thick *evergreen* lvs. and black berries, pine barrens of N.J. and s. [00]**S. wálteri** Pursh., Redberried Greenbrier, has bright red berries, N.J. s. and w.

FIG. 24. *Smilax rotundifolia*, Common Greenbrier. Brts. and lf. nat. size; cross section of stem × 6.

CLASS II. DICOTYLEDONS

The remaining families in the book belong to this group of the angiosperms, characterized by two cotyledons (seed leaves) in the embryo, and, in the stem, a ring of vascular bundles surrounding the pith. In the woody Dicots, the cambium ring develops from this ring of vascular bundles and forms annual layers of wood and bark.

SALICÂCEAE — WILLOW FAMILY

Dioecious; both staminate and pistillate fls. in catkins; lvs. alternate (in *S. purpurea* often opposite), simple; bark bitter; wood light and soft.

SÀLIX — WILLOW

Buds covered by a *single hollow-conical* scale, often more or less flattened against brts.; true terminal bud absent; lvs. mostly long and rather narrow. (Pl. III. and Figs. 25, 26, 27, 28).

To recognize a willow at any season is easy: this is one of the few native genera in our area that have a single cap-like scale covering the bud. But to determine the sp. is quite another matter. Some spp. are very variable, especially nos. 7, 8, 10, and 11, and there is also considerable hybridization among the different spp. However, the large tree willows, represented by the first 6 spp. below, are not difficult. Of the shrubs, *discolor*, *cordata* and *sericea* are the commonest, the first very variable, but with the underside of lf. glaucous and the teeth irregular; the second with the lvs. cordate (at base only) and with stipules usually prominent and persistent; the third, usually in swamps or wet places, with the lf. underside *silky*. *S. lucida* is a very distinct sp., not so common, but easily recognized by its very long-pointed, shining lvs.

Large trees.
 Brts. long and pendulous.
  Brts. yellow; lvs. usually slightly silky beneath.........3. *S. alba* var. *tristis* 77
  Brts. olive-green or brownish; lvs. long (3–6 in.) narrow, finely toothed, glabrous
   beneath (Pl. III, 1)....................................5. *S. babylonica* 77
 Brts. not as above.
  Brts. brittle at base; lvs. oblong-lanceolate.
   Lvs. linear-lanceolate, often scythe-shaped, serrulate, green beneath (Fig.
    25)...1. *S. nigra* 77
   Lvs. oblong-lanceolate (2½–6 in.), long-pointed, serrate, glabrous and blue-
    green beneath; petioles glandular (Pl. III, 2)...............2. *S. fragilis* 77
  Brts. not markedly brittle at base (although occasionally some hybrids have
   this character to some extent).
   Lvs. (1½–5 in.) rather long-pointed, shining above, cult....6. *S. pentandra* 77
   Lvs. not as above.
    Brts. olive-brown; lvs. serrulate, glaucous and silky beneath....3. *S. alba* 77
    Brts. yellow; lvs. as above but smoother (Pl. III, 8).4. *S. alba* var. *vitellina* 77
Shrubs (Nos. 8, 9, 11, 13 sometimes small trees).
 Mature lvs. glabrous beneath.
  Lvs., at least some of them, opposite or nearly so, glaucous beneath, often
   nearly entire (Pl. III, 3) ....14. *S. purpurea* 78
  Lvs. all alternate.
   Petiole glandular where it joins lf. blade; lvs. long-pointed, bright green and
    shining (Fig. 28)......................................13. *S. lucida* 78
   Petiole not glandular.
    Lvs. irregularly serrate, glaucous beneath, rarely somewhat pubescent
     (Fig. 26, rt.).......................................8. *S. discolor* 77

Lvs. sharply and regularly serrate; stipules usually prominent and often
persistent; brts. velvety (Pl. III, 7)....................7. *S. cordata*   77
Mature lvs. pubescent beneath.
Cult. "Pussies," large and early like those of *S. discolor* but lvs. shorter, ½–¾
as broad as long...9. *S. caprea*   77

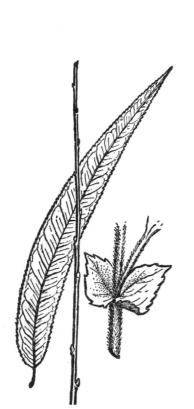
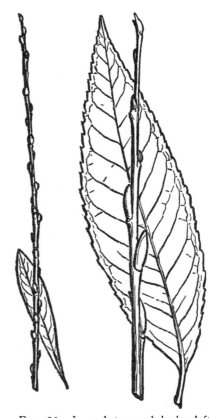

FIG. 25. *Salix nigra*, Black Willow.
Lf. and brt. nat. size; stipules × 4.

FIG. 26. Lvs., brts., and buds: left,
Salix tristis, Dwarf Pussy Willow; rt., *S.
discolor*, Pussy W.; nat. size.

Native.
Pubescence of appressed white silky hairs; wet places (Pl. III, 4)12. *S. sericea*   78
Pubescence not as above.
Lvs. nearly or quite entire (sometimes undulate-dentate in *S. humilis*).
Lvs. linear-lanceolate, comparatively short, the margins revolute (Pl.
III, 5)..15. *S. candida*   78
Lvs. oblanceolate, margin often irregular but not definitely toothed
(Pl. III, 6)...10. *S. humilis*   77
Lvs. serrate, oblong to oblong-ovate; veins much sunken on upper sur-
face (Fig. 27)....................................11. *S. bebbiana*   77

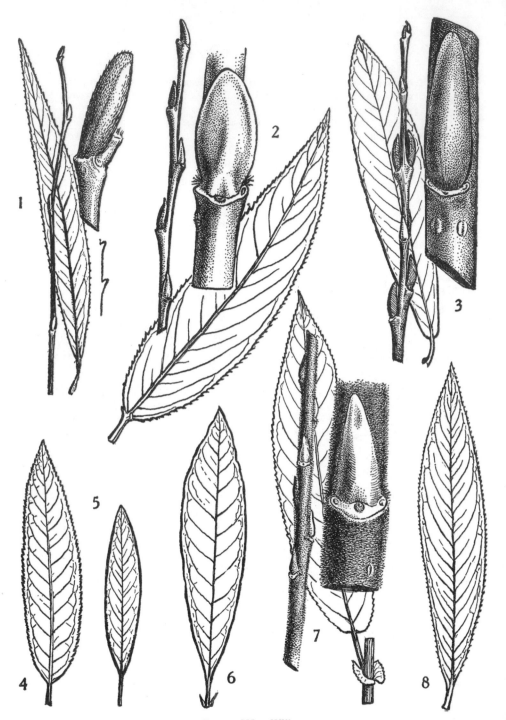

PLATE III. Willows.

1. *Salix babylonica*, Babylon Weeping W.   Lf. and brt. nat. size; bud × 10; lf. teeth × 5.   2. *S. fragilis*, Crack W.   Lf. and brt. nat. size; bud and brt. × 5.   3. *S. purpurea*, Purple-osier W. Lf. and brt. nat. size; bud and brt. × 5.   4. *S. sericea*, Silky W.   Lf. nat. size.   5. *S. candida*, Sage W., Hoary W.   Lf. nat. size (somewhat shorter than average).   6. *S. humilis*, Prairie W. Lf. nat. size.   7. *S. cordata*, Heartleaf W.   Lf., stipules and brt. nat. size; bud and lf. scar × 5. 8. *S. alba* var. *vitellina*, Yellowstem White W.   Lf. nat. size.

1. **S. nìgra** Marsh. Black W. (Fig. 25). Large tree; brts. brittle at base; lvs. linear-lanceolate, running out to a long fine point, often scythe-shaped, green and smooth beneath, except sometimes pubescent on veins. N.B. to Ala. and w.

2. **S. frágilis** L. Crack W. (Pl. III, 2). Large tree; lvs. very long-pointed, light green or blue-green and at length smooth beneath, serrate, with glands at base of blade; brs. and brts. *very brittle, easily knocked off with the finger.*

3. **S. álba** L. White W. Large tree; lvs. lanceolate, long-pointed, finely toothed, *silky pubescent* and whitened beneath, in one var. silky pubescent also above; brts. greenish. Rather rare. Eu. A weeping form, var. *tristis*, is sometimes cult. and resembles no. 5, but has yellow brts.

4. **S. álba** L. var. **vitellìna** (L.) Stokes. Yellowstem White W. (Pl. III, 8). Large tree, more common than the last. Mature lvs. similar, but smooth or nearly so, whitish beneath. Brts. yellow. (*Vitellina* means egg-yellow). Frequently natzd.

5. **S. babylónica** L. Babylon Weeping W. (Pl. III, 1). May be recognized by its *very long, slender, olive-brown, drooping* brts. and narrow lvs. paler and smooth beneath. Eurasia. Often cult.

6. S. pentándra L. Bay-leaved W. (Laurel W. SPN). Tree to 60 ft. Lvs. rather long-pointed, shining, green on both sides; petioles glandular. Sometimes cult. Eu. Fine specimens in Public Garden, Boston.

7. ⁰S. cordàta Muhl. Heartleaf W. (Pl. III, 7). Shrub with finely serrate lvs., green beneath and cordate (at least some of them) at base; stipules prominent and *long-persistent.* N.B. to Va. and w. (*S. rigida* Muhl.).

8. **S. díscolor** Muhl. Pussy W. (Fig. 26, rt.). Usually a shrub; fl. buds much larger than lf. buds and opening early; lvs. smooth (sometimes slightly pubescent), and glaucous beneath, irregularly serrate. Lab. to Md. and w.

9. S. cáprea L. Goat W. Closely related to **S. discolor** but more ornamental; often cult. Can be distinguished by large "pussies" (1–1½ in. long) and by pubescent brts. (until autumn) and rugose lvs., pubescent beneath. Eurasia.

10. ⁰S. hùmilis Marsh. Prairie W. (Pl. III, 6). Small, low shrub; lvs. oblanceolate or oblong-lanceolate, tomentose beneath; buds and brts. hairy. Nfd. to N.C. and w.

The Dwarf Pussy W., ⁰S. tristis Ait., (Fig. 26, left) is closely related to *S. humilis* (*S. humilis* var. *microphylla* (Anderss.) Fern.) but has much smaller lvs. Me. to Fla. and w.

11. **S. bebbiàna** Sarg. Bebb W. (Fig. 27). Large shrub or small tree; lvs. obovate to elliptic-lanceolate, usually pubescent beneath,

with conspicuous veins somewhat sunken below upper lf. surface.   Fr. has long beak.   Nfd. to Pa. and w.   (*S. rostrata* Richards.).

12. ⁰**S. serícea** Marsh.   Silky W. (Pl. III, 4).   Shrub, 6–12 ft. high; lvs. narrow, finely serrate, *silky beneath*.   Very common in swamps, and growing in large colonies.   Me. to Va. and w.

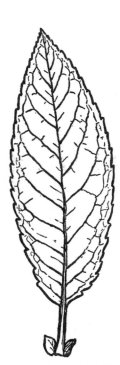

FIG. 27.   *S. bebbiana*, Bebb Willow, nat. size.

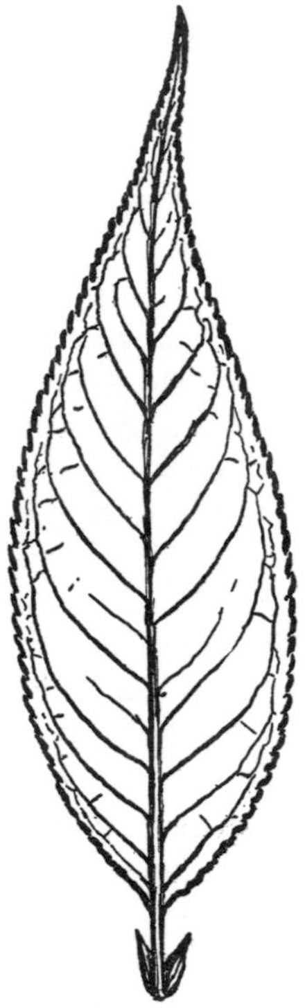

FIG. 28.   *S. lucida*, Shining Willow, nat. size.

13. ⁰**S. lùcida** Muhl. (Fig. 28).   Shining W.   Lvs. very long-pointed and *shining;* shrub of moist soil.   Lab. to Md. and w.

14. ⁰**S. purpùrea** L. (Pl. III, 3).   Purple-osier W.   Tall shrub, brts. smooth, purplish when young, becoming gray, lvs. oblanceolate, toothed toward apex, pale beneath; *buds (and lvs.) often opposite*.   Eu. and N.Afr. to C.Asia and Japan.   Sometimes natzd.   Ornamental and used for finer basket work.

15. ⁰**S. cándida** Fluegge.   Hoary W. (Pl. III, 5) has brts. and underside of lvs. white-woolly.   Lvs. entire, with revolute margins.   Lab. to Pa. and w.   Also called Sage W. (SPN).

Pópulus — Poplar

Buds with *many* scales; true terminal bud and stipule scars present; bark very bitter; pith 5-angled; lvs. in general broader than in *Salix;* bark on young trees and brs. smooth and pale, gray or yellowish, but dark and rough on old trunks. Often called Cottonwood, from the fine cottony hairs attached to the seeds, especially marked in no. 4. The Aspens have much flattened petioles, causing the lvs. to flutter in the slightest breeze.

Petioles flattened, at least toward lf. blade.
  Lvs. ovate to rounded.
    Lvs. with small rounded teeth (crenate-serrate); buds slender (fl. buds larger), appressed to brt. and shining as if varnished (Pl. IV, 3)...2. *P. tremuloides* 79
    Lvs. with larger, sharper teeth; buds slightly white-downy (Pl. IV, 4)
                                               3. *P. grandidentata* 81
  Lvs. more or less triangular.
    Cult.
      Lvs. comparatively small, as wide, or wider than long; tree of columnar habit (Pl. IV, 2)....................................5. *P. nigra* var. *italica* 81
      Lvs. larger, about 2½–3 in. wide, reddish and ciliate when young, long-pointed; much planted along city streets (Pl. V, 4)
                                               8. *P. canadensis* var. *eugenei* 81
    Native. Lvs. larger, delta-shaped; buds containing much resin (Pl. V, I and Fig. 29)...4. *P. deltoides* 81
Petioles rounded.
  Buds with abundant, fragrant resin; lvs. with metallic luster beneath.
    Lvs. heart-shaped, ciliate (Pl. IV, 1).......................7. *P. candicans* 81
    Lvs. ovate or ovate-lanceolate, minutely ciliate (Pl. V, 2)...6. *P. balsamifera* 81
  Buds and lvs. not as above.
    Native. Buds large (½–¾ in. long); rare sp. of swamps and bottom lands.
                                               9. *P. heterophylla* 81
    Natzd. Buds smaller (terminal about ¼ in. long); brts. and lvs. more or less covered with a white down; lvs. small with 3–5 large teeth or small lobes. (Pl. V, 3)...1. *P. alba* 79

1. **P. álba** L. White P. (Pl. V, 3). Young brts. and under surface of lvs. with a *white, felty covering;* lvs. rhombic with very coarse teeth or almost lobes; buds more or less woolly; young bark very pale gray or nearly white; dark, rough bark appears later than in other spp. of *Populus.* Often natzd. Eurasia. Several vars. are cult. of which *P. a.* var. *pyramidàlis* Bunge. with columnar form and lvs. more deeply toothed, or lobed is occasionally seen. Latter also known as *P. bolleàna* Lauche.

2. **P. tremuloìdes** Michx. Quaking Aspen (Pl. IV, 3). Common in forests, usually in rather dry locations. Lvs. ovate to rounded, with small, regular, usually rounded teeth; petioles flattened, making lvs. tremble with slightest breeze; lf. buds slender, very sharp-pointed, shining

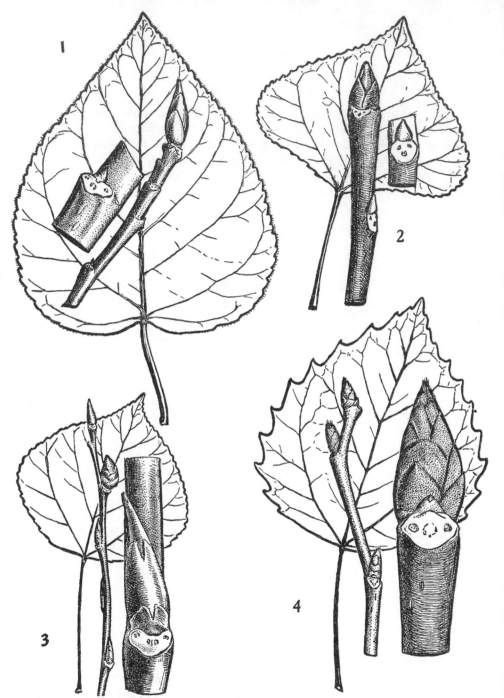

PLATE IV. Poplars.

1. *Pòpulus candicans*, Balm-of-Gilead P.   Lf. about nat. size.   Terminal bud and brt. nat. size; part of brt. with lat. bud and lf. scar × 4.

2. *P. nigra* var. *italica*, Lombardy P.   Lf. nat. size; brts. with terminal and lat. buds somewhat enlarged.

3. *P. tremuloides*, Quaking Aspen, Trembling A.   Lf. and brt. nat. size; large bud near tip, a fl. bud; part of brt. showing lat. bud and lf. scar × 5.

4. *P. grandidentata*, Bigtooth Aspen.   Lf. and brt. nat. size, terminal bud and lf. scar × 5.

as if varnished, appressed close to brt.; young bark gray- or yellow-green; old bark nearly black. Lab. to Pa. and w.

3. **P. grandidentàta** Michx. Bigtooth Aspen (Pl. IV, 4 and Fig. 1). Lvs. round-ovate, with large, irregular teeth; buds plumper than in the last, somewhat downy; young bark usually more distinctly yellow than that of the last. Common in forests, N.S. to N.C. and w. Very young trees or shoots of this sp. often have another leaf form—large and very woolly beneath.

4. **P. deltoìdes** Marsh. Eastern Cottonwood, Necklace Poplar (Pl. V, 1 and Fig. 29). Lvs. deltoid in shape, margins more or less ciliate when young; brts. yellowish, often showing 3 ridges extending downward below lf. scars; buds large, smooth, containing a yellow, fragrant resin; young bark much like that of the last; old bark grayish. Often cult. on account of its rapid growth. Que. to Fla. and w.

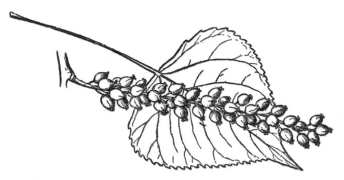

Fig. 29. *Populus deltoides*, Cottonwood, Necklace-Poplar. Fr. and lf. × ½.

5. P. nìgra L. var. itálica Muenchh. Lombardy P. (Pl. IV, 2). Horticultural var. much cult. and easily recognized by its erect, columnar habit; lvs. comparatively small, usually wider than long.

6. **P. balsamífera** L. Balsam P., Tacamahaca (Pl. V, 2). Buds large, containing fragrant resin; lvs. ovate, acuminate, minutely ciliate, rounded at base, smooth with metallic luster on pale lower surface; petioles rounded. Lab. to N. E., N.Y. and w.

7. **P. cándicans** Ait., Balm-of-Gilead P. (Pl. IV, 1), with large, sticky, fragrant buds, and with brts., petioles, and under sides of broadly ovate lvs. (especially on the veins) pubescent, or at least the margins of the lvs. ciliate, is also occasionally seen. Often planted and believed to be an infertile hybrid of nos. 4 and 6. 8. The Carolina P., *P. canadénsis* var. *eugènei* Schelle, (Pl. V, 4), is said to be a var. of a cross between *P. deltoides* and *P. nigra*. Much planted along city streets. Similar forms with a different parentage probably also occur. 9. **P. heterophýlla** L., Swamp Cottonwood, is (in our area) a rare sp. in wooded swamps and borders of ponds, with large lvs., woolly beneath, at least when young. Conn. to Fla. and w.

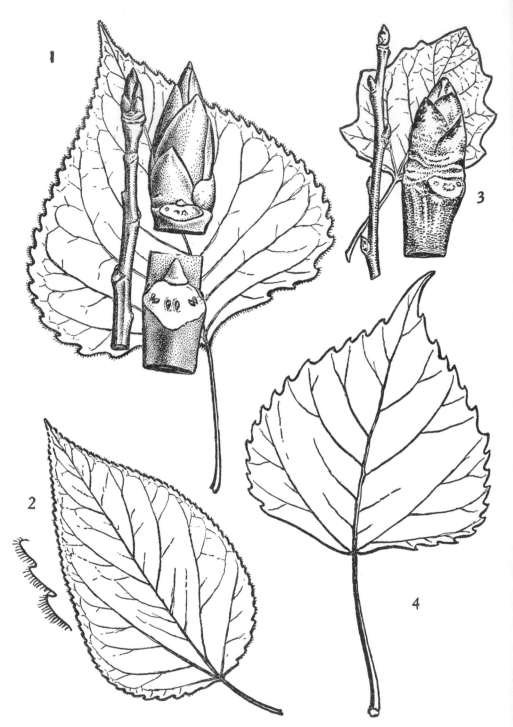

PLATE V. Poplars concl.

1. *P. deltoides*, Cottonwood, Necklace P.   Lf. and brt. nat. size; bud and part of brt. × 4.
2. *P. balsamifera*, Tacamahac P. (Balsam P.)   Lf. nat. size; teeth and cilia × 5.
3. *P. alba*, White P., Abele.   Lf. and brt. nat. size; terminal bud × 5.   (Lf. below average size.)
4. *P. canadensis* var. *eugenei*, Carolina P.   Lf. nat. size.   (Sparsely ciliate when young.)

MYRICÀCEAE — SWEET GALE FAMILY
MYRÌCA

1. ⁰**M. pensylvánica** Lois. Northern Bayberry (Pl. VI, 1). Lvs. aromatic, falling late; brts. and buds with long black hairs (seen under lens) and dotted with yellow resin glands; *bud globose*, about 1/12 in. in diam.; staminate catkins enclosed in the winter buds (left brt. in fig.); fr. grayish white, waxy. Nfd. to w. N.Y. and N.C. chiefly near seacoast.

Toward the south of our range, i. e. in s. N.J., the northern Bayberry passes into another sp., 2. the Southern Bayberry, **Myrìca cerífera** L., although the two exist side by side for a time. The Southern Bayberry is very similar, but a large plant, sometimes 35 ft. high, with *pointed evergreen* lvs., and occurs from N.J. to Fla. and Tex.

3. ⁰**M. gàle** L. Sweet Gale (Pl. VI, 2). A usually larger shrub than **M. pensylvánica,** with similar lvs., but with resin-dotted fr. in *cone-like bunches* at the ends of the brts., and with usually pointed buds. Borders of ponds and in swamps. Lab. to N.E. & N.Y. and w. In mts. to N.C. and Tenn. Also Eurasia.

COMPTÒNIA — SWEET-FERN

⁰**C. peregrìna** (L.) Coult. Sweet-fern (Pl. VI, 3), with fragrant, more or less deeply cleft, fern-like lvs., is common in sterile sandy soil. N.S. to N.C. and w. In both of these last 2 spp. the staminate catkins are more or less exposed and prominent during the winter (see figs.), and are fairly erect. (*Myrica asplenifòlia.*)

JUGLANDÀCEAE — WALNUT FAMILY

Lvs. of hickory and walnut, our two genera in this family, pinnately compound; true terminal bud present; fr. a nut with 2 outer coverings, an outer husk and an inner shell; large forest trees. In the walnut the outer husk does not split on ripening; in the hickory it splits into 4 parts.

JÙGLANS — WALNUT

Husk of fr. not splitting when ripe; pith chambered; buds naked (at least without typical scales); lf. scars shield-shaped or 3-lobed; stipule scars lacking; lfts. numerous (11–23).

1. **J. cinèrea** L. Butternut, White W. (Pl. VI, 4). *Transverse, downy pads above triangular lf. scars;* buds gray-brown; pith dark brown; nuts long; lfts. 11–17; bark with broad, light, smooth, lengthwise stripes. N.B. to Ga. and w.

2. **J. nìgra** L. Black W. (Eastern Black W. SPN) (Pl. VI, 5). Lf. scars heart-shaped; buds grayish; pith lighter brown; nuts spherical; lfts.

PLATE VI.   Sweet Gale Family and Walnuts.

1. *Myrica pensylvanica*, Northern Bayberry.   Lf. and flowering (left) and vegetative (rt.) brts. nat. size; end of twig showing resin globules × 4.

2. *Myrica gale*, Sweet Gale.   Brt. and lf. nat. size; bud containing staminate catkin × 5.

3. *Comptonia peregrina*, Sweet-fern.   Brt. and lf. nat. size; lat. bud × 10.

4. *Juglans cinerea*, Butternut.   Brt. nat. size; lf. × ¼; lat. (superposed) buds and lf. scar × 4.

5. *Juglans nigra*, Black Walnut.   Brt. nat. size; lf. × ¼; lat. (superposed) buds and lf. scars × 4.

13–23; bark dark and rough, without stripes. W. Mass. to Fla. and w. Often cult. for its nuts and shade. A handsome tree.

3. The Persian W., *J. règia* L., the so-called "English W.," but native from southeastern Eu. to India and China, is sometimes cult. and forms are hardy in the Northeastern States. Lfts. 5–9, *entire* or nearly so, bark silvery gray, long remaining smooth. Very variable in size and shape of nut and in thickness of shell; now cult. on a commercial scale in southern Calif. and parts of Ore. and Wash.

4. The Siebold W., *J. sieboldiàna*, is occasionally cult. Resembles **J. cinerea** closely but brs., buds, lvs., and fr. are more vigorous, the last in clusters of 12–20. Japan.

CÁRYA (HICÒRIA) — HICKORY

Husk splitting open into 4 valves, at least part of the way down from apex, when ripe; pith *not chambered;* buds with imbricated scales (valvate in *C. cordiformis*); lvs. pinnately compound; lf. scars shield-shaped or 3-lobed; stipule scars lacking; lfts. (5–11) usually fewer than in *Juglans.* Forms of *C. glabra* and *C. tomentosa* sometimes occur, and are difficult to determine with exactness; *C. ovata* and *C. cordiformis* are quite distinct.

Buds yellow, scales valvate, nuts thin shelled (Pl. VII, 4)........6. *C. cordiformis* 87
Buds not yellow, scales imbricate, nuts thick shelled.
  Buds small (¼-⅝ in. long); brts. slender; husk of fr. thin.
    Fr. obovoid, husk not splitting to base, kernel astringent (Pl. VII, 3)
          4. *C. glabra* 87
    Fr. ovoid, husk splitting at length to base, kernel sweet..........5. *C. ovalis* 87
  Buds, at least the terminal, larger (¾–1 in. long); brts. stout; fr. husk thick.
    Lvs. and brts. tomentose; bark close; husk not splitting readily to base (Pl. VII, 2)................................3. *C. tomentosa* 85
    Lvs. and brts. smooth or slightly pubescent; bark shaggy; husk splitting to base.
      Lfts. 7–9; young brts. pale orange......................2. *C. laciniosa* 85
      Lfts. usually 5, the terminal one larger; young brts. red-brown (Pl. VII, 1)
          1. *C. ovata* 85

1. **C. ovàta** (Mill.) K. Koch. Shagbark H. (Pl. VII, 1). Bark shaggy; buds of medium size, with *outermost scales (at least in some buds) produced into long points; husk of fr. thick, splitting readily to base; shell of nut white,* kernel sweet, the hickory nut of commerce; lvs. downy below when young, later usually smooth, or slightly pubescent, with usually 5 lfts. (sometimes 7), the terminal one much larger. Que. to Minn. s. to Fla. and Tex.

2. **C. laciniòsa** (Michx.) Loud. Shellbark H. Brts. *pale orange;* lfts. usually 7–9, husk and shell of nut thick; nut very large, sometimes more than 2 in. long. Otherwise like *C. ovàta,* but more western in its range. N.Y. to Neb., s. to Tenn. and Okla.

3. **C. tomentòsa** Nutt. Mockernut H. (Pl. VII, 2). Bark close, i.e., not readily peeled off with the fingers; *buds large,* the terminal one ½–¾ in. long, outer scales early deciduous; *brts. stout and thick,* usually pubescent; husk of fr. thick, not splitting to base; nut light brown, shell thick;

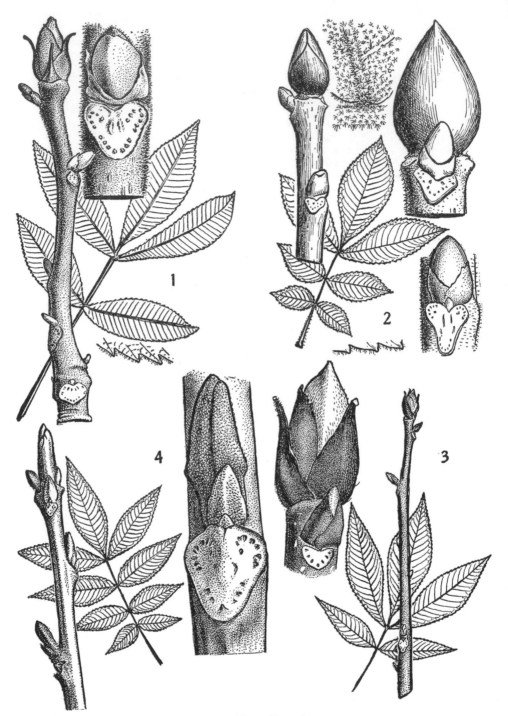

PLATE VII. Hickories.

1. *Carya ovata*, Shagbark Hickory. Brt. nat. size; lf. × ¼; lat. bud and lf. scar × 4; teeth of lfts. × 5.

2. *C. tomentosa*, Mockernut H. Brt. nat. size; lf. × ¼; stellate hairs from lf. underside enlarged; terminal and lat. buds × 2; at bottom, lf. teeth enlarged × 5.

3. *C. glabra*, Pignut H. Brt. nat. size; lf. × ¼;terminal bud × 5.

4. *C. cordiformis*, Bitternut H. Brt. nat. size; lf. × ¼; lat. buds (superposed) and lf. scar × 4.

lvs. pubescent below, with stellate hairs (see fig.), with 5–7, often 9 lfts., very fragrant when crushed. Sometimes called "Bigbud H." Mass. to Ont. w. and s.

4. **C. glàbra** (Mill.) Sweet. Pignut Hickory (Pl. VII, 3). Bark scaly; buds smaller (⅛–½ in. long), outer scales early deciduous; *brts. slender, smooth;* fr. ovate, husk thin, not splitting more than halfway to base; lvs. smooth, with usually 5 lfts. (may have 3–7), smaller than in the last. Mass. to Ont. w. and s.

Another sp. of pignut, the Small Pignut, 5. **C. ovàlis** (Wang.) Sarg., is also recognized, characterized by small (about 1 in. long), ovoid fr. with a *thin, warty husk,* which splits tardily to near the base, and by bark which is often shaggy on old trunks; lfts. 5–7; both lfts. and brts. scurfy-pubescent while young, glabrous when mature. Mass. to Wis. and s. Several vars. of this sp. depending on the size and shape of the fr., are described. Here belongs *Hicòria microcàrpa,* Britton. Mass. to Ga. and w.

6. **C. cordifórmis** (Wang.) K. Koch. Bitternut H. (Pl. VII, 4). Bark close; *buds sulphur colored, bud scales valvate; husk and shell of nut thin* (can be indented with the fingernail); kernel bitter; lfts. *numerous* (7–11). Closely related to *C. illinoensis,* the pecan nut of the southern States. Que. to Minn., Fla. and Tex.

BETULÀCEAE — BIRCH FAMILY

Monoecious; both staminate and pistillate fls. in catkins, except in *Corylus,* where the pistillate fls. are in short, few-flowered heads; staminate catkins *naked throughout the winter in all genera* except *Carpinus;* pistillate catkins naked throughout the winter *only* in *Alnus* (except in *A. crispa*), in other genera enclosed in the buds. The naked catkins are a conspicuous winter character, and are therefore an easy means of identification of membership in this family. Lvs. simple, alternate, usually doubly serrate.

BÉTULA — BIRCH

Staminate catkins conspicuous in winter on mature trees, pendulous; bark marked by horizontally elongated lenticels; dwarf shoots numerous along 2-year-old or older twigs; true terminal bud lacking except at tips of the numerous dwarf shoots; buds with 2–3, sometimes more, scales showing; stipule scars narrow; frs. in compact cylindrical spikes, each individual fr. (a winged nut) subtended by a 3-lobed bract. (See Birches of the N.Y. City Region. Series 11, Bull. 10, Publ. by Audubon Soc. N.Y.C.)

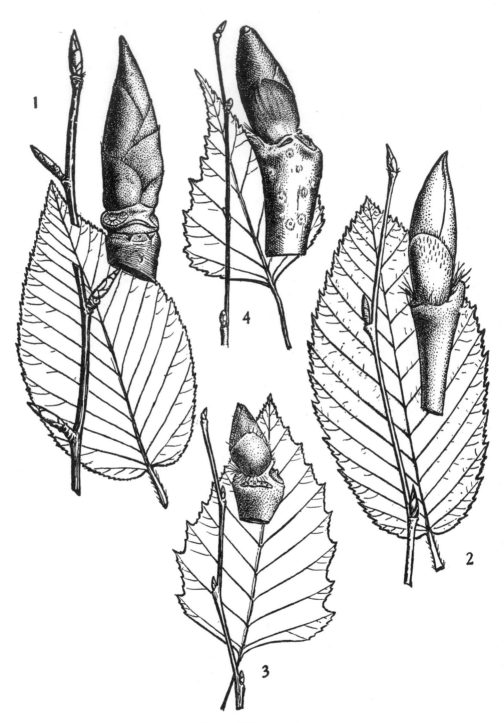

PLATE VIII. Birches.

1. *Betula lenta*, Sweet B. Lf. and brt. nat. size; bud × 5.
2. *B. alleghaniensis*, Yellow B. Lf. and brt. nat. size; bud × 5.
3. *B. nigra*, River B. Lf. and brt. nat. size; bud × 5.
4. *B. populifolia*, Gray B. Lf. and brt. nat. size; bud × 5.

(This key does not include no. 6, B. pumila,
a dwarf sp. of bogs and swamps. (Pl. IX, 2.)

Brts. with aromatic, wintergreen taste.
  Bark of trunk dark brown, not curly; brts. brown and smooth, with strong
    wintergreen taste (Pl. VIII, 1)...............................1. *B. lenta* 89
  Bark of trunk dull yellow, more or less curly in young trees; young brts. with
    yellowish tint, somewhat pubescent (Pl. VIII, 2).........2. *B. alleghaniensis* 89
Brts. without wintergreen taste.
  Bark of trunk white.
    Cult. Eu. spp., (formerly *Betula alba*). Bark often peeling off in thin strips.
      Brts. finely pubescent; lvs. pubescent beneath, at least when young (Pl.
        IX, 1)..8. *B. pubescens* 91
      Brts. glabrous, dotted with resin glands; lvs. gummy when young, smooth;
        brs. pendulous.......................................7. *B. pendula* 91
    Native.
      Bark of trunk chalky white, separating easily into thin papery layers; lvs.
        ovate, wedge-shaped or rounded at base; in var. *cordifolia* more or less
        cordate at base (Pl. IX, 3).......................5. *B. papyrifera* 91
      Bark of trunk dull white, not easily separating into thin layers; lvs. truncate
        or cuneate at base, more or less triangular, long-pointed. Common small
        tree, often in clumps, on poor soil (Pl. VIII, 4).........4. *B. populifolia* 89
  Bark of trunk reddish, ragged, with papery, curling, irregularly arranged strips;
    lvs. more or less wedge-shaped at base, beneath whitish and when young
    pubescent (Pl. VIII, 3)...................................3. *B. nigra* 89

1. **B. lénta** L. Sweet B., Black B. (Pl. VIII, 1). Bark dark brown to
black; twigs with strong wintergreen flavor; buds sharp-pointed and long
($\frac{1}{4}$–$\frac{1}{2}$ in.); fr. bracts *smooth*. Has finely and doubly serrate lvs. like
Ostrya and *Carpinus* with which it may be confused when young; but it
can always be distinguished by the taste of the nearly smooth brts. Sw.
Me., to Md. and upland Ga. and Tenn.

2. **B. alleghaniensis** Britton. Yellow B. (Pl. VIII, 2). Bark peeling
into thin, silvery yellow, often curling, ribbon-like layers; twigs with
wintergreen flavor (but not as strong as in the last), pubescent, at least
when young; lvs. like the last; spikes of fr. a little thicker and shorter;
fr. bracts *pubescent*. Sometimes in moist soil, but also frequent at higher
elevations. Nfd. to Ia. s. to Md., upland to N.C. and Tenn. and w.

3. **B. nìgra** L. River B., Red B. (Pl. VIII, 3). Bark papery, curly,
and cinnamon red; this and the irregularly doubly serrate lvs., often
almost rhombic, bluish beneath, and its preference for river and stream
banks, shores of ponds and swamps, are its most characteristic features.
Se. N.H. and ne. Mass. through sw. Conn. to Fla. and w. Good specimens along Hutchinson R. Pkwy., Westchester Co., N.Y. near junction
with Merritt Pkwy.; also along Bronx R. Pkwy., n. of N.Y. Bot. Garden.

4. **B. populifòlia** Marsh. Gray B. (Pl. VIII, 4). To 30 ft. Bark
dirty white, not chalky, nor dividing readily into thin layers; lvs. triangu-

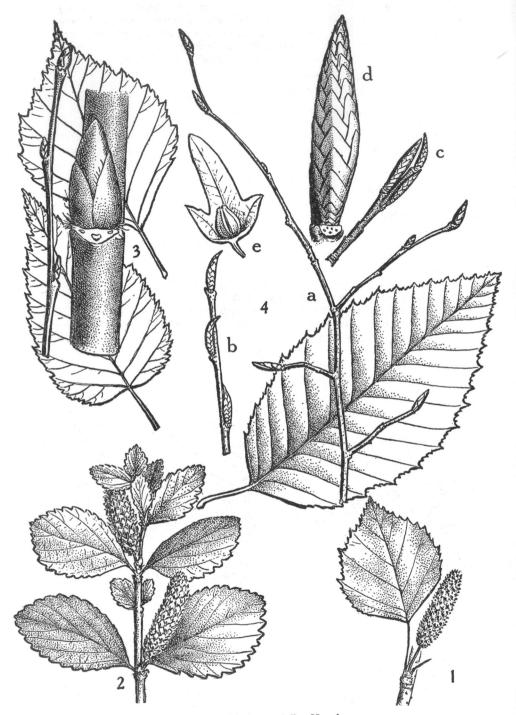

PLATE IX. Birches and Eu. Hornbeam.

1. *Betula pubescens*, Downy Eu. White B.   Lf. and fruiting catkin nat. size.
2. *B. pumila*, Low B.   Brt. with lvs. and fruiting catkins nat. size.
3. *B. papyrifera*, Paper B.   Two forms of lvs. and brt. nat. size; part of brt. with bud and l.f
scar × 5.
4. *Carpinus betulus*, Eu. Hornbeam.   a, lf. with brt. nat. size;   b, part of brt. showing curved
buds;   c, tip of brt. × 2;   d, uppermost lat. bud enlarged;   e, fruiting bract and fr. nat. size.

lar, *very long-pointed;* buds short, ⅛ to ¼ in.; brts. roughened with resin dots (see fig.); a small tree, typically with several oblique trunks from a single base. Very common, usually in poor soil. P.E.I. and Ont. to Del., uplands of Va. and w.

5. **B. papyrífera** Marsh. Paper B., Canoe B. (Pl. IX, 3). To 90 ft. or more. Similar to the Gray Birch but the *lvs.* are *ovate,* heart-shaped [var. *cordifolia* (Reg.) Fernald], or wedge-shaped at base; brts. more or less pubescent; *bark chalky white,* separable into thin layers; and the tree is larger and taller with usually a single main trunk. Typically a northern sp. Lab. to s. N.E. & N.Y. to upland Pa. and Va. and w.

6. ⁰**B. pùmila.** L. Low B. (Pl. IX, 2). Dwarf sp. (to 9 ft.) of bogs and swamps with small, rounded, toothed lvs., and brts. pubescent when young. Nfd. to N.J. and w.

7. **B. péndula** Roth (*B. verrucòsa* Ehrh.). Eu. Weeping White B. The common, cult. White Birch. Bark rather dirty or sometimes creamy white; brts. with resin glands, not pubescent; lvs. rhombic-ovate. The cut-leaved horticultural var. *gracilis* Rehd. is the one most commonly seen.

8. **B. pubéscens** Ehrh. Downy Eu. White B. (Pl. IX, 1). Can be distinguished from the last by the glandless, pubescent brts. and the more erect brs. Commonly cult.[1]

ÁLNUS — ALDER

With the exception noted below, buds are *stalked* (Pl. X, 1, 2) with 2 or sometimes 3 nearly equal, valvate scales; both staminate and pistillate catkins are naked and conspicuous during the winter; old fr. heads (conelike) also persistent. But in the Green Alder, *A. crispa* (Pl. X, 3), chiefly a northern sp., buds are *sessile,* covered by 3–6 unequal, imbricated scales, and the pistillate catkins are enclosed in the buds during the winter.

Buds sessile, covered by 3–6 unequal, imbricated scales. Pistillate catkins enclosed
  during winter in scaly buds; nutlets winged (Pl. X, 3)............3. *A. crispa* 92
Buds stalked, covered by 2 or 3 nearly equal valvate scales; both pistillate and
  staminate catkins naked, nutlets without wings.
  Flowering in spring.
    Lvs. finely tomentose or glaucous beneath (Pl. X, 1)...........1. *A. rugosa* 92
    Lvs. green, smooth or sometimes finely pubescent beneath.
      Lvs. finely serrulate, not gummy, native sp. (Pl. X, 2).....2. *A. serrulata* 92
      Lvs. more coarsely toothed, young growth very gummy, Eu. sp. (Fig. 30)
                                                          4. *A. glutinosa* 92
  Flowering in autumn. Native of Del. and Md................5. *A. maritima* 94

[1] The complex of Amer. and Eu. White Birches is most interesting. Our Paper B. has its counterpart in the Eu. sp. *B. pubescens,* in fact our tree was formerly called simply a variety of the Eu. sp. (then *B. alba*). Our Gray B. has its corresponding sp. in Eu., called *B. pendula,* which resembles it in many ways. Here may be a case of parallel evolution.

1. ⁰**A. rugosa** (Du Roi) Spreng. Speckled A. (Pl. X, 1). Lvs. ovate, downy and more or less glaucous beneath, doubly serrate; bark speckled with large lenticels; *pistillate catkins recurved.* Often grows in drier soil than the next.

2. ⁰**A. serrulata** (Ait.) Willd. Smooth or Common A. (Pl. X, 2). Lvs. green on both sides, almost regularly serrate, narrowing somewhat at their bases; *pistillate catkins erect.* Usually in moist soil.

The above two alders are the common spp. and often grow in the same habitat, but are usually easily distinguished, although intermediate forms sometimes occur.

Fig. 30. *Alnus glutinosa*, Eu. Alder. Lf. nat. size.

3. ⁰**A. crispa** (Ait.) Pursh. Green or Mountain A. (Pl. X, 3). Buds *not stalked;* lvs. round-oval, irregularly serrulate; pistillate catkins enclosed in scaly buds and therefore not apparent in winter. The alder of cool shores and the mts. in n. N.E. and N.Y. to w. N.C. and w. Var, **mollis** Fernald, with lvs. pubescent beneath, is the common form in the Me. coast region.

4. A. glutinòsa (L.) Gaertn. Eu. A. (Fig. 30). Large tree to 75 ft. of Eurasia and N. Afr., occasionally escaped in our area. Lvs. very gummy when young, rounded or emarginate at apex, coarsely and doubly dentate.

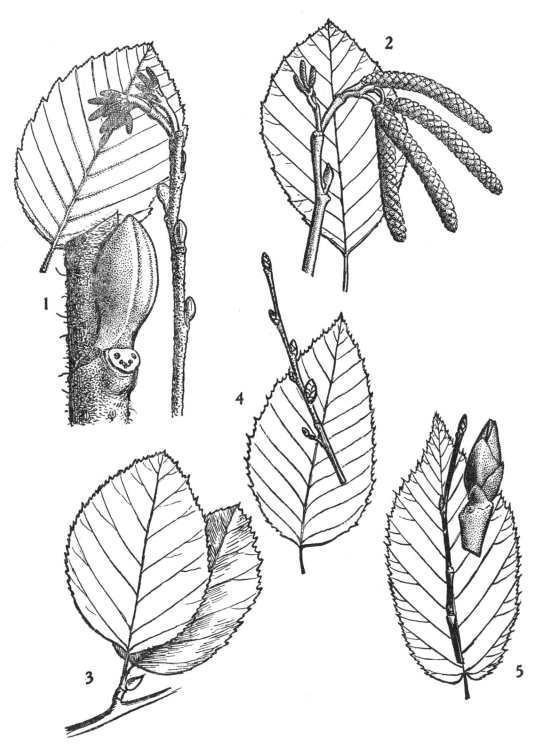

PLATE X. Alders, Amer. Hornbeam, and Hop-hornbeam. 1. *Alnus rugosa*, Speckled A. Lf. and brt. with pistillate catkins nat. size; bud, lf. scar and portion of brt. × 5. 2. *A. serrulata*, Common or Smooth A. Lf. and brt. with both pistillate and staminate catkins nat. size. 3. *A. crispa*, Green A. Lvs. and brt. nat. size. 4. *Carpinus caroliniana*, Amer. Hornbeam. Lf. and brt. nat. size. 5. *Ostrya virginiana*, Amer. Hop-hornbeam. Lf. and brt. nat. size; bud × 5.

5. The Seaside A., **A. marítima** (Marsh.), is a shrub or small tree.   Buds acute, scurfy pubescent; lvs. elliptic to oblong, cuneate at base.   *Blooms in fall.*   Del. and Md. and in Okla.

CARPÌNUS — HORNBEAM

Neither staminate nor pistillate catkins evident during the winter, both being enclosed in the buds; buds small, pointed, angular, with 10–12 scales exposed in 4 rows; true terminal bud lacking; stipule scars present; frs. in loose, pendulous clusters, each individual nut-like fr. subtended by a 3-lobed bract; bark smooth, fluted (with "muscle-like" ridges).

1. **C. caroliniàna** Walt. Amer. Hornbeam, Blue-beech (Fig. 31).   Bark smooth, steel-gray. Likely to be confused with *Ostrya* when young, but the lvs. are entirely smooth above, often reddish in fall, and the buds are smaller, reddish, usually *angled*, and show more scales; also, fl. catkins are absent in fall and winter.   N.E. to Fla. and w.   (Pl. X, 4).

Fig. 31. *Carpinus caroliniana*, Amer. Hornbeam.   Fr. brt. nat. size.

2. C. bétulus L.   Eu. Hornbeam, Eu. Blue-beech (Pl. IX, 4).   A larger tree (60–70 ft.) than the Amer. sp. (30–40 ft.); bark similar; lvs. thicker, with veins *sunken* in upper surface, turning yellow in fall, and often persistent all winter; bracts large, 1½ in. long (in Amer., to 1 in. long), margin of bract *nearly entire;* buds longer (¼ in.) than in Amer. ($^1/_6$ in.).   Eu. to Iran.

ÓSTRYA — HOP-HORNBEAM

O. virginiàna (Mill.) K. Koch.   Hop-hornbeam (Pl. X, 5).   Small tree, with bark in long, narrow, loose, ragged, vertical strips; buds ovoid, generally *tinged with green*, about *6 striate* scales exposed; true terminal bud lacking; staminate catkins, *often in 2's or 3's*, present, at least on mature trees; lvs. oblong-ovate, sharp-pointed, slightly hairy on both sides; frs. enclosed in bladder-like sacs which *occur in cone- or hop-like clusters.*   N.S. to Fla. and w.

The Hornbeam and the Hop-hornbeam are more often confused than any other native trees.   When they are large trees the difference in the bark is a clear distinction; also the staminate catkins of the Hop-hornbeam, like little fingers, in 2's or 3's scattered here and there in the crown.   But in very young trees the difference is not so clear, since the lvs. are very much alike.   Then one must examine carefully the buds for the characters given above.   Especially noteworthy is the angular, more scaly character of the Hornbeam bud.

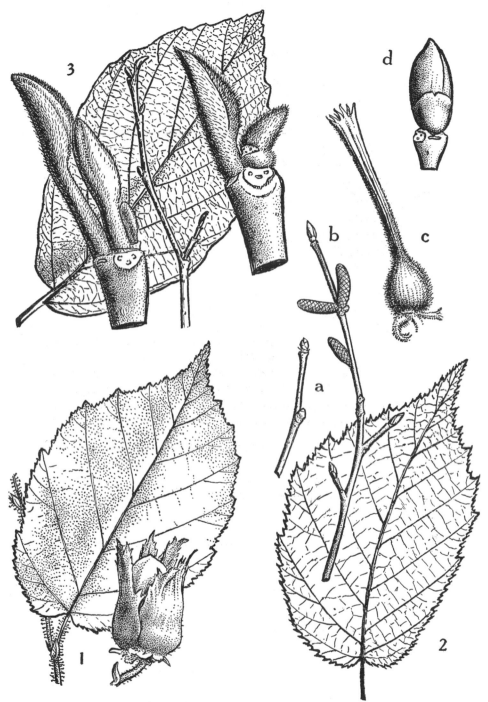

PLATE XI. Hazels and Witch-hazel.

1. *Corylus americana*, Hazelnut. Lf. and brt. nat. size. The fr. is that of the Eu. Hazel or
   Filbert.
2. *C. cornuta*, Beaked H. Lf. and brt. b, with catkins nat. size; a, tip of rt. showing glabrous
   character; c, fr. nat. size; d, uppermost lat. bud, with lf. scar at base (left), stipule scar
   showing at base in front, and end of season's growth (not evident) at rt. × 5.
3. *Hamamelis virginiana*, Witch-hazel. Lf. and brt. nat. size; tips of 2 brts. × 5, showing
   (left) terminal and lat. buds, lf. scar and stipule scar; (rt.) terminal bud and last lat. bud.

Córylus — Hazelnut

Shrubs; staminate catkins *gray*, pendulous; buds somewhat flattened, blunt, with 4–6 scales exposed; true terminal bud lacking; lvs. often in 3 ranks. Stipule scars present.

1. ⁰**C. americàna** Walt. Hazelnut (Amer. Filbert SPN) (Pl. XI, 1). Brts. and petioles with stiff, glandular hairs; lvs. oval, pointed, downy below; fr. enclosed in a broad leafy involucre. Me. to Sask., s. to Ga. and w.

2. ⁰**C. cornùta** Marsh. Beaked H. (Pl. XI, 2). Brts. more slender than in the last, smooth or only slightly hairy; lvs. broader than in the last; fr. with a long, tube-like involucre. Nfd. to B.C., s. to Ga. and w.

3. ⁰Corylus avellàna L., Eu. Hazelnut, known in the trade as the Filbert (Pl. XI, 1, fr.) has a slightly larger nut than the Amer. sp. Much cult. for its nuts and also in ornamental vars. such as the Cutleaf, Golden, Purple, and Weeping F. A somewhat larger shrub than *C. americana*, with glandular pubescent lvs. and involucre only slightly longer than the nut, while in *C. americana* the involucre is about twice as long as the nut.

Most of the Hazelnuts are shrubs, but *C. colúrna*, the Turkish H., is a large tree, to 75 ft. A fine, tall specimen is in the Oxford (Eng.) Botanic Garden and another at the Arnold Arboretum.

In our area *C. americana* is the common sp. *C. cornuta*, in rich rocky woods, can be readily distinguished in fruit by the cylindrical involucre, and at any time of the yr. by the smooth, more slender brts. without the stiff hairs of *C. americana*. Wild nuts of either sp. are usually hard to get — squirrels.

FAGÀCEAE — BEECH FAMILY

Mainly trees (a few are shrubs) with alternate, simple, straight-veined lvs.; monoecious; characterized particularly by the fr., a nut, more or less surrounded by a woody or spiny involucre = the cup of the acorn, or the bur of the chestnut or beech (Pls. XII, XIII, XIV).

Fàgus — Beech

Easily recognized by the *very long* (sometimes nearly 1 in.), *narrow, cylindrical, sharp-pointed, scaly buds;* stipule scars linear, nearly meeting around brt.; true terminal bud present; bark light gray and *smooth*, even in old trees; in winter the pale, dead lvs. tend to persist on the tree; nuts triangular, sweet and edible, usually in a spiny bur, the involucre. Only one sp. (*F. grandifolia*) is native in the U.S. The Eu. sp., *F. sylvatica*, and its vars., are commonly cult. Both spp. are beautiful shade trees.

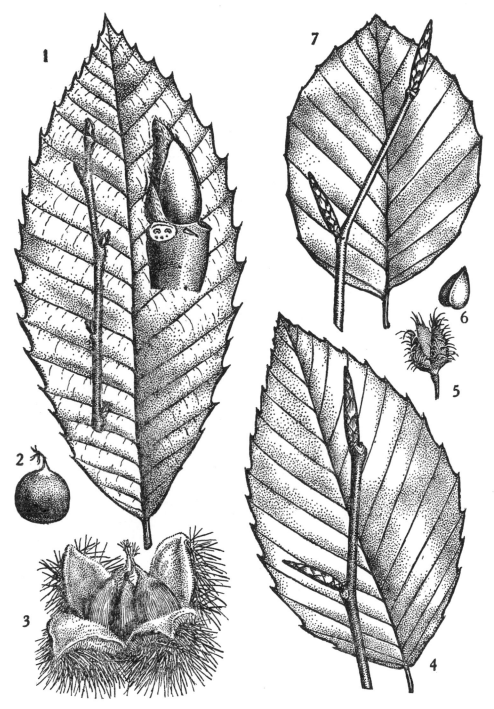

PLATE XII. Beech and Chestnut.

Castanea dentata, Amer. Chestnut. 1. Lf. and brt. nat. size; uppermost lat. bud, lf. scar, stipule
    scar, and stub of end of this year's growth (at rt.) × 8. 2. Nut about nat. size. 3. Bur
    or involucre with nuts about nat. size.
Fagus grandifolia, Amer. Beech. 4. Lf. and brt. 5. Involucre. 6. Nut. All nat. size.
Fagus sylvatica, Eu. Beech. 7. Lf. and brt. nat. size.

1. **F. grandifòlia** Ehrh. Amer. Beech (Pl. XII, 4, 5, 6). Lvs. ovate-oblong, coarsely serrate, 2½–5 in. long, with 9–14 pairs of veins. P.E.I. to Fla. and w. Var. **caroliniana** has darker, thicker lvs., with shorter teeth and *blunt* buds. S.e. Mass. to Fla. and w. (Naushon and Marthas Vineyard Is.)

2. **F. sylvática** L. Eu. Beech (Pl. XII, 7). Lvs. similar, but smaller (2–4 in.), with only 5–9 pairs of veins, and with smaller teeth than in the Amer. sp.; bark somewhat darker than in the latter. Cent. and s. Eu.

Vars. of the Eu. B. most commonly planted are:

F. sylvática L. var. purpùrea Ait. Copper or Purple B. Lvs. copper-colored or purple, changing to dark green in late summer.

F. sylvática L. var. incìsa Hort. Cut-leaf Eu. B. (Fern-lvd. B.). Lvs. deeply and variously cleft and toothed. Var. *asplenifòlia* Duchartre is similar but with very narrow lvs.

F. sylvática L. var. péndula Loud. Weeping Eu. B. Brs. drooping. A famous specimen, perhaps the largest in the U.S., is in Flushing, N.Y. (*Brooklyn Bot. Gard. Record*, **15**: 44, 45, and figs. 4 and 5. 1926).

CASTÀNEA — CHESTNUT

Characterized particularly by the fr. consisting of 1–3 (rarely 5–7) rounded nuts, surrounded by a prickly bur, the involucre, which splits at maturity into usually 4 valves. Botanically the chestnuts fall into 2 groups: chestnuts proper with usually 2 or 3 nuts in the bur, and the chinkapins or bush chestnuts with usually only 1 nut in the bur (see no. 2). There are several spp. of chinkapins in the southern U.S. and also in China. Buds blunt, ovoid, light to dark brown, with only 2 or 3 scales showing, true terminal frequently lacking; pith *star-shaped;* on erect shoots lvs. usually in 5 ranks, on lateral shoots in 2 ranks. The Amer. sp. is now rare as a tree, having been killed by a parasitic fungus brought into the U.S. from the Orient; but young shoots are often seen growing from the old roots, which have persisted because of the greater resistance of the roots to the disease.

Large, old trees of chestnut now living in our area are either of the Chinese or Japanese sp. which are blight-resistant, the former very much so. When not in leaf (when the glands on the under side of the Jap. readily distinguish it) the twig characters are a ready means of distinction, the Chinese having shoots of the yr. *buff yellow* and more or less *downy*, while the Jap. has a *brown* or *red-brown* twig, usually glabrous (sometimes with a few hairs). The Eu. or Spanish C., *C. sativa*, the nut now sold in the markets, also known as the Italian C., is not usually hardy in our area, and also very susceptible to blight.

1. **C. dentàta** (Marsh.) Borkh. Amer. C. (Pl. XII, 1, 2, 3). Lvs. resembling those of the Amer. Beech, but longer, wider, and with larger, sharper teeth, *smooth beneath;* brts. and buds smooth. Central Me. to Mich., s. to Fla. and w.

2. **C. pùmila** Mill., Allegheny Chinkapin. Spreading shrub or small tree with smaller lvs. *white-tomentose* beneath, and *one* nut, about the size of a small pea, in the bur; brts. and buds downy. Appears in southern part of our area. N.J. and Pa. s. and w.

3. **C. mollíssima** Bl. Chinese C. Medium sized, usually spreading tree, to 60 ft., recognizable especially by the downy, buff-colored shoots of the yr., and by the thick, shining lvs., whitish-tomentose beneath, at least when young. Now being much cult.; blight resistant to a high degree. Nuts large, often 1 in. across, variable in flavor, sometimes sweet but often tasteless until cured. China, Korea; sometimes natzd. in the neighborhood of plantations in Amer.

4. **C. crenàta** Sieb. and Zucc. Japanese C. Small tree to 30 ft. or more. Lvs. thinner than in last, and dull, the teeth often reduced to bristles, tomentose when young, smoother with age; but peppered with tiny glands on the under surface. Shoots of yr. brown, glabrous, or finely pubescent. Nuts very variable, some large, 2 in. in diam., some medium, and some small. More commonly cult. in our area than any other sp., sometimes locally natzd., and occasionally very blight resistant.

Quércus — Oak

Characterized by a cluster of buds toward the end of the brt., the bud scales being numerous and in 5 ranks; by simple lvs., lobed or cleft in the majority of the spp.; and by the fr., known as the acorn.[1] Here the involucre (the cup of the acorn) does not usually entirely enclose the fr. as in the beech and chestnut. Stipule scars small; pith star-shaped in section. Although there are many subdivisions, for convenience 2 main groups of oaks may be recognized in this region, as follows:

White Oaks	Black Oaks
(Including Chestnut Oaks)	(Including Willow Oaks)
Leaf lobes or teeth rounded, at least without bristle tips (sharp-pointed in nos. 6, 8).	(See nos. 20–23 incl.) Leaf lobes with bristle tips.
Buds blunt (except in nos. 7, 8).	Buds more or less sharp-pointed.
Acorns mature first yr. (except in Turkey O.).	Acorns mature second yr.[2]

[1] For illustrations of acorns, not shown here, see "Common oak trees in summer" by the author. Bull. 10, Ser. 10, School Nature League, N.Y.C. Publ. by Natl. Audubon Soc., N.Y.C. 1940.

[2] On this account all spp. of the black oak group, if old enough to bear fr., will be found in any autumn to have 2 sizes of nuts: those of the current yr., still tiny, and those in their second yr. now ripening.

SUMMER KEY

(Not including nos. 4, 9, 18, 22, 23 which occur only in the extreme southern part of our area, but see no. 18).

Lvs. entire or nearly so, i.e. not lobed or distinctly toothed.
  WILLOW OAKS. See note 1 for no. 20, p. 109.
    Lvs. narrowly lanceolate, glabrous, about ½ in. wide (Pl. XIII, 3)..20. *Q. phellos* 109
    Lvs. broader, 1-2 in. wide, pubescent beneath, shining above. .21. *Q. imbricaria* 109
Lvs. not entire; either lobed or toothed.
  Lvs. lobed.
    Lobes rounded and not bristle-tipped. WHITE OAKS.
      Some of lvs. with small ear-like lobes at base, obovate in general outline, and
        comparatively small (2–4 in.); acorns stalked; cult. (Pl. XIV, 4)
                                                           10. *Q. robur* 106
      None of lvs. with ear-like lobes at base.
        Lvs. small (2–5 in. long), cult.
          Buds with thread-like scales intermixed with regular scales (Pl. XIV, 3)
                                                           12. *Q. cerris* 106
          Buds without thread-like scales; acorns sessile..........11. *Q. petraea* 106
        Lvs. averaging larger (4–10 in. long); native.
          Brts. glabrous, at least when mature.
            Lvs. glabrous beneath, at least when mature (Pl. XIV, 1)..1. *Q. alba* 104
            Lvs. pubescent beneath.
              Lobes usually quite short, becoming rounded teeth near tip of lf.
                (Pl. XIV, 2)................................5. *Q. bicolor* 104
              Lobes longer, especially toward base of lf.; lobing often divided into
                an upper and a lower section by a pair of deep, wide sinuses
                (Pl. XIV, 8)..............................3. *Q. macrocarpa* 104
          Brts. pubescent; lvs. pubescent beneath, typically 5-lobed, the middle
            pair expanded at tips, producing, with the top lobe or lobes, an effect
            like a Latin cross (Pl. XIV, 6).....................2. *Q. stellata* 104
    Lobes not rounded but bristle-tipped, sometimes apparently rounded in nos.
      16, 19, but then tipped with bristles. BLACK OAKS.
      Shrub or small tree; lvs. white-downy beneath, very variable in shape; in
        poor soil; common on rocky ridges and sand barrens (Pl. XIII, 4)
                                                           17. *Q. ilicifolia* 108
      Large trees (except sometimes in no. 19).
        Lvs. pubescent beneath.
          Lvs. not deeply lobed, broadly inversely triangular, sometimes scarcely
            lobed, but bristle tips evident (Pl. XIII, 6)......19. *Q. marilandica* 109
          Lvs. more or less deeply lobed, typically obovate in general outline;
            lobes usually 7, broader near tip than in Red O.; inner living bark
            orange, outer bark *black and rough.* Several lf. forms common, one
            (shown in fig.) in which the lobing is not so pronounced (Pl. XIII, 5)
                                                           16. *Q. velutina* 107
        Lvs. smooth beneath when mature (sometimes, especially in no. 15, short
          tufts of hairs in axils of veins beneath).
          Lvs. deeply cleft, i. e., lobes long.
            Lateral lobes standing out more or less at rt. angles to midrib.

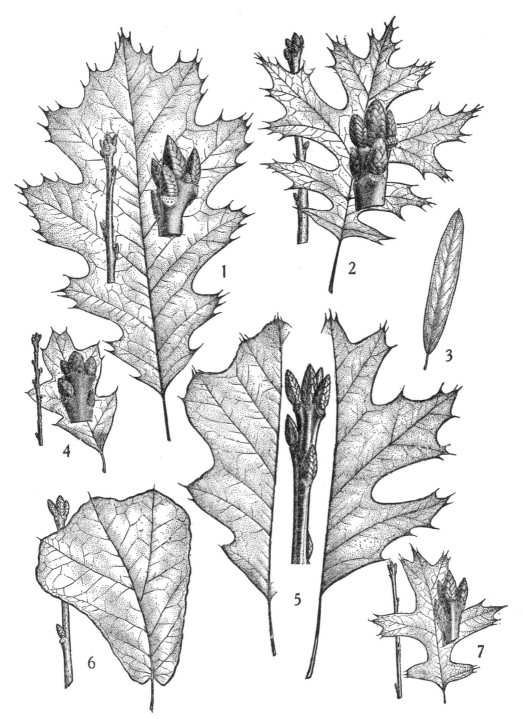

PLATE XIII. Black Oaks.

1. *Quercus rubra*, Red O.   Lf. and brt. ½ nat. size; terminal bud cluster × 2½.
2. *Q. coccinea*, Scarlet O.   Lf. and brt. ½ nat. size; terminal and bud cluster × 2½.
3. *Q. phellos*, Willow O.   Lf. ½ nat. size.
4. *Q. ilicifolia*, Bear O.   Lf. and brt. ½ nat. size; terminal bud cluster × 2½.
5. *Q. velutina*, Black O.   Sides: 2 types of lvs. ½ nat. size; center: brt. and buds × 2½.
6. *Q. marilandica*, Blackjack O.   Lf. and brt. with buds × ½.
7. *Q. palustris*, Pin O.   Lf. and brt. ½ nat. size; terminal bud cluster × 2½.

Lvs. small, usually 3–4 in. long; acorns small, cups very shallow and saucer-shaped; short pin-like brts. plentiful; often in wet soil (Pl. XIII, 7)............................14. *Q. palustris* 107

Lvs. averaging larger (3–6 in.) turning scarlet in fall; acorn cup covering about ½ of acorn, inner living bark gray, sometimes with pink tinge (Pl. XIII, 2)......................15. *Q. coccinea* 107

Lateral lobes not standing out at rt. angles to midrib, lobes many (7–11) more or less triangular in outline; acorns large, cups saucer-shaped, ¾ in. or more in diam. (Pl. XIII, 1)..........13. *Q. rubra* 107

Lvs. not deeply cleft; lobes 7–11, and more or less triangular in outline; acorns large, cups saucer-shaped (Pl. XIII, 1)..........13. *Q. rubra* 107

Lvs. toothed, the teeth without bristle-tips, although often with short points (except no. 12, where the teeth are often so pronounced that they might be classed as lobes). Chestnut Oaks.

Lvs. elliptic or oblong.

Shrub. Lvs. with 3–7 pairs of teeth, grayish and more or less stellate-pubescent beneath. Rocky ridges and sand barrens (Pl. XIV, 5)..6. *Q. prinoides* 106

Trees.

Cult. in parks and private grounds. Lvs. with 4–9 pairs of teeth, the teeth often lobe-like and themselves toothed. Winter buds containing a few awl-shaped scales. Native of se. Eu. and Asia Minor (Pl. XIV, 3)
12. *Q. cerris* 106

Native. Lvs. with many fairly sharp teeth, much like those of the native chestnut, whitish-pubescent beneath. Usually on limestone soil. (Fig. 32)....................................8. *Q. muehlenbergi* 106

Lvs. obovate or ovate.

Lvs. without teeth toward base, distinctly pubescent beneath; bark light gray-brown, sloughing off on *young* brs.; acorn on long stalk; moist soil (Pl. XIV, 2)..5. *Q. bicolor* 104

Lvs. toothed to or nearly to base; nearly or quite glabrous and pale beneath; bark dark brown, not scaly, with deep, V-shaped grooves, not exfoliating on young brs.; acorn cup thin, hemispherical; usually in rocky places but sometimes sandy soil (Staten Island, N.Y.) (Pl. XIV, 7).....7. *Q. prinus* 106

WINTER KEY

Based on buds, brts., bark and acorns. The lvs., being durable, will be found on the ground all winter and in some spp. remain on the tree until spring. This key does not include nos. 4, 9, 18, 22, 23, which occur only in the southern part of our area; but see no. 18.

Buds sharp-pointed.

Buds large, especially the terminal ones, ⅜–½ in. long.

Buds pubescent.

Buds with yellow-brown or grayish pubescence, usually strongly 5-sided; bark of trunk rough, nearly black; lvs. usually 7-lobed; large tree (Pl. XIII, 5)..16. *Q. velutina* 107

Buds red-brown, rusty-hairy, angled; bark of trunk nearly black and divided into nearly square plates; lvs. broadly inversely triangular, often scarcely lobed; small shrubby tree to 30 ft. (Pl. XIII, 6)......19. *Q. marilandica* 109

Buds glabrous (often pubescent at tip in no. 13).

Buds dark brownish-red, usually shining and somewhat 5-sided; bark of old
    trunks not deeply grooved (Pl. XIII, 1)..................13. *Q. rubra* 107
Buds light brown or yellow, bud scales often ciliate; bark of old trunks with
    deep, V-shaped grooves (Pl. XIV, 7)...................7. *Q. prinus* 106
Buds smaller (¼ in. long or less).
    Numerous short, pin-like brts. at nearly rt. angles to main brs.; buds glabrous,
        rather slender; brts. glabrous and shining (Pl. XIII, 7)...14. *Q. palustris* 107
    Pin-like brts. not evident or at least not numerous.
        Brts. pubescent or puberulous; shrubby; lvs. of Black O. type, very variable
            (Pl. XIII, 4).......................................17. *Q. ilicifolia* 108
        Brts. glabrous.
            Buds glabrous.
                Buds shining; bud scales with pale, often jagged or ciliate margins; lvs.
                    of Willow O. type, not lobed.................21. *Q. imbricaria* 109
                Buds dull; bud scales with membranaceous margins; lvs. of Chestnut O.
                    type, toothed (Fig. 32)......................8. *Q. muehlenbergi* 106
                Buds pubescent in upper half, lower half smooth and reddish; lvs. turning
                    scarlet in fall (Pl. XIII, 2)........................15. *Q. coccinea* 107
Buds blunt; although they may come to a definite point, it is more obtuse than in the
    above group.
    Awl-shaped or thread-like scales evident in the buds (see Pl. XIV, 3, 8), usually
        not numerous in no. 5.
        Bark of trunk light-colored, scaly.
            Young brts. with exfoliating bark; acorns on stalks an inch or more long;
                buds pubescent in upper half (Pl. XIV, 2)............5. *Q. bicolor* 104
            Young brts. not as above; buds covered with pale pubescence; acorns with or
                without stalks, cup of acorn fringed at rim (Pl. XIV, 8)..3. *Q. macrocarpa* 104
        Bark of trunk dark colored, furrowed, not scaly; acorns cylindrical, more than
            1 in. long, covered to ½ by cup-shaped cup with awl-shaped, recurved scales
            (Pl. XIV, 3)...12. *Q. corris* 106
    Awl-shaped scales not present in the buds, or at least not conspicuous.
        Buds, at least the terminal ones, large, about ¼ in. long, sometimes slightly
            less in weak brts.
            Brts. pubescent or puberulent.
                Brts. stout, light colored; lvs. with squarish lobes (Pl. XIV, 6)..2. *Q. stellata* 104
                Brts. slender, dark colored; lvs. with various outlines (Pl. XIII, 4)
                                        17. *Q. ilicifolia* 108
            Brts. glabrous.
                Bark of trunk light colored.
                  Brts. yellow or yellow-brown; outer bark on younger parts of large brs.
                    exfoliating (Pl. XIV, 2)............................5. *Q. bicolor* 104
                Brts. reddish or glaucous, or often with bronzy tint; outer bark on
                  younger parts of large brs. intact (Pl. XIV, 1)............1. *Q. alba* 104
             Bark of trunk dark colored, furrowed.
                Acorns stalked; lvs. often with small ear-like lobes at base (Pl. XIV, 4)
                                      10. *Q. robur* 106
              Acorns sessile; lvs. without ear-like lobes.............11. *Q. petraea* 106
Buds small, less than ¼ in. long.
    Shrub with Chestnut O. type of lf. (Pl. XIV, 5)............6. *Q. prinoides* 106
    Large tree with Willow O. type of lf. (Pl. XIII, 3).........20. *Q. phellos* 109

Species in the White Oak Group

1. Q. álba L.   White Oak (Pl. XIV, 1).   Buds *smooth and blunt;*
bark light gray and scaly; brts. smooth, usually reddish-brown at maturity, often glaucous; acorns ovoid-
oblong, cups shallow; lvs. deeply
lobed, sometimes nearly to midrib,
pale and smooth below; dead lvs.
often persist on tree in winter. Var.
latilòba Sarg. is a common form:
in this the lf. lobes are broad, and
extend usually less than halfway
to the midrib.   The form figured is
very close to this var.   Que. to Fla.
and w.

2. Q. stellàta Wang.   Post O.
(Pl. XIV, 6).   Buds pubescent and
blunt; brts. stout, tomentose; bark
of trunk red-brown, or sometimes
lighter, scaly; acorns small, with
hemispherical cups; lvs. *with squarish lobes*, stellate pubescent beneath; small tree.   Common on
L.I. and Hunter's I., N.Y.   Se.
Mass. to Fla. and w.

3. Q. macrocárpa Michx.   Bur
O. (Pl. XIV, 8).   Large tree; bark
light-colored, scaly; large acorns;
cups with fringed borders; lvs. often separated into 2 parts (upper
and lower) by a pair of deep sinuses
½ or ⅔ down lf.   Often cult. N.B.
to Pa., s. and w.

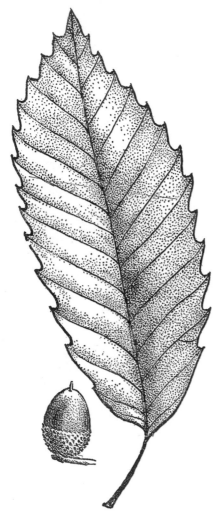

4. **Q. lyràta** Walt.   Overcup O.   Lvs.
resembling those of **Q. alba** but lobes are
fewer and usually in 2 sets (somewhat as
in **Q. macrocarpa**) separated by a deep
cleft, the lower lobes smaller, lvs. white
tomentose or sometimes green and pubescent beneath.   Cup often nearly encloses

Fig. 32. *Quercus muehlenbergi,* Yellow
Oak, Chinkapin Oak (SPN).   Lf. and
acorn nat. size.

the acorn, hence common name.   Recorded from s. N.J., (Riddleton, Salem Co.)
and Md. to Fla. and w.

5. Q. bìcolor Willd. (*Q. platanoides* Sudw.) (Pl. XIV, 2).   Swamp-
White O.   Buds blunt, with fine capillary scales often present in the

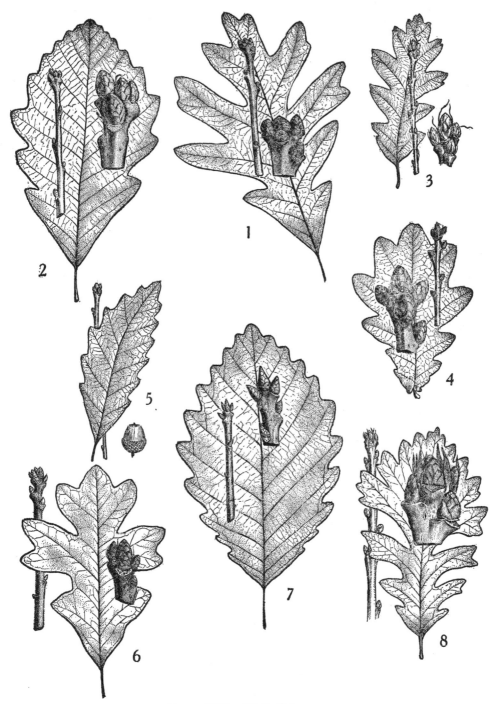

PLATE XIV. White Oaks.

1. *Quercus alba*, White O. Lf. and brt. ½ nat. size; terminal bud cluster × 2½.
2. *Q. bicolor*, Swamp-White O. Lf. and brt. ½ nat. size; terminal bud cluster × 2½.
3. *Q. cerris*, Turkey O. Lf. and brt. ½ nat. size; terminal bud cluster × 2½.
4. *Q. robur*, English O. Lf. and brt. ½ nat. size; terminal bud cluster × 2½.
5. *Q. prinoides*, Dwarf Chinkapin O. Lf., brt. and fr. ½ nat. size.
6. *Q. stellata*, Post O. Lf. and brt. ½ nat. size; terminal bud cluster × 2½.
7. *Q. prinus*, Chestnut O. Lf. and brt. ½ nat. size; terminal bud cluster × 2½.
8. *Q. macrocarpa*, Bur O. Lf. and brt. ½ nat. size; terminal bud cluster × 2½.

terminal bud cluster; brts. at length smooth and yellowish; lvs. narrower and entire towards the base, tomentose beneath, with numerous rounded lobes, but these usually not so deep as in **Q. alba** (rarely lobed halfway to midrib); *bark exfoliating on young portions of brs.* Moist soil. Que. to Ga. and w.

6. **Q. prinoìdes** Willd. (Pl. XIV, 5). Dwarf Chinkapin O. A dwarf sp.; brts. brittle, slender, smooth; buds small, very blunt; lvs. of the Chestnut O. type, regularly crenately toothed, gray-downy beneath. Vt. and Mass. to Ala. and w.

7. **Q. prìnus** L. Chestnut O. (*Q. montana* Willd.) (Pl. XIV, 7). Buds yellowish, smooth, *pointed;* bark not flaky, dark brown, in mature trees with *deep, more or less vertical grooves,* v-shaped in cross section; *cup* somewhat warty, but *thin,* covering $\frac{1}{3}$–$\frac{1}{2}$ of acorn; lvs. with numerous, regular, rounded teeth. Sw. Me. to Ga. and w.

> **Q. prinus** is the common Chestnut O. in our area, inhabiting *dry, rocky ridges* (sandy soil on S.I., N.Y.) and characterized especially by its *thick, deeply grooved bark* and by its lvs., pale and minutely downy beneath, which resemble those of the Amer. Chestnut, but the teeth are *rounded.* A much rarer sp. is 8. **Q. muehlenbérgi** Engelm., Chinkapin O. (SPN), Yellow O. (Fig. 32), which prefers *limestone ridges* and *ledges,* and has narrower lvs., rather sharply toothed, whitish tomentulose beneath, and thin *flaky bark.* Nw. Vt. and n. N.Y., s. to Fla. and w. (Also known as *Q. acuminata* Sarg.). A third sp. of this Chestnut O. group, 9. **Q. michaùxi** Nutt., the Swamp Chestnut O., also called the Basket O. is more southern; N.J. and Del. to Fla. and w. Easily distinguished by its *silvery-white, scaly bark.* Lvs. much like those of **Q. prinus** but usually tomentose beneath. Occurs in *moist soil, borders of streams, swamps and bottom lands.*

10. Q. ròbur L. English O. (Pl. XIV, 4). Brts. glaucous, usually reddish above, green beneath; buds short, thick, blunt. Resembles *Q. alba* in some respects, but lvs. are smaller, with 3–7 pairs of lobes, often with basal lobes auricled. The acorns, borne on stalks 1–3 in. long, ripen the first yr. and although the bark is dark colored, this sp. is closely related to our White O.

> 11. *Q. petràea* Lieblein, Durmast O. is the other sp. of oak native in Britain. Distinguished (1) by sessile acorns (also known as *Q. sessiliflora*) (2) by lvs. cuneate at base and (3) by lvs. with more lobes (5-9 pairs). Usually a smaller tree, not so commonly cult. as *Q. robur.*

12. Q. cérris L. Turkey O. (Pl. XIV, 3). Capillary scales of buds numerous; lvs. with shallow, abruptly pointed lobes; acorn (about 1 in. long) enclosed about $\frac{1}{2}$ by the cup, ripening the second yr.; bark dark colored and rough. Native se. Eu. and w. Asia. Occasionally cult. Several vars.

Species in the Black Oak Group

13. Q. rùbra L. Red O. (Pl. XIII, 1). Buds large, those at tip of brt. about ¼ in. long or more, red- or dark-brown and usually shining; rounded on the sides (not strongly angled), somewhat pubescent towards the pointed tip. Lvs. with numerous bristle-tipped lobes which are more or less triangular in rough outline; *smooth beneath* except for occasional tufts of hairs in axils of veins. Brts. usually strongly ridged or fluted. Acorns large, about 1 in. long, with flat saucer-shaped cups. The bark of old trunks is comparatively smooth, vertically but not deeply grooved, and with smooth, light-colored strips between the grooves. The Red O. is a common tree with us, especially in rocky situations; it is a vigorous, fast grower, and is best recognized by the dark red, shining buds, the many-lobed (7–11) lvs., the sinuses *varying considerably as to depth*, and the *large acorns*. P.E.I. to Ga. and w. Var. **boreàlis** (Michx f.) Farw. Northern Red O. has *deeper cups*, enclosing about ⅓ of acorn and young bark paler gray and smoother. Que. to nw. Pa. in mts. to N.C. and w.

14. Q. palústris Muenchh. Pin O. (Pl. XIII, 7). Buds about ⅛ in. long, sharp-pointed, angled, entirely smooth and usually shining, brown; *many short, pin-like brs. throughout tree;* lower main brs., especially in younger trees, droop characteristically downward; lvs. small and deeply cleft, the lobes often standing out nearly at right angles to the long axis of the lf.; *acorns small*, about ½ in. in diam., often striped, nearly hemispherical, with flat cups. Best recognized by its smooth, sharp-pointed buds, pin-like brs., and small acorns with flat cups. Prefers moist soil, but sometimes occurs in dry sandy soil. Much planted as a street tree and in parks and estates. Mass. to S.C. and w.

15. Q. coccínea Muenchh. Scarlet O. (Pl. XIII, 2). Buds ⅛–¼ in. long, fairly smooth and reddish brown at base; gray and distinctly pubescent above middle, and pointed at tip. Lvs. *scarlet in the fall*, hence the name; shaped as in Pin O., deeply cleft, but larger; lobes less numerous than in Red O. Acorn with hemispherical cup and white meat. *Inner, living bark not yellow*, but reddish or pinkish; outer bark black and rough. Apt to be confused with the Pin O., but lacks the pin-like brts., and its buds are larger and pubescent in the upper half. Also may be mistaken for the Black O. Me. to Ga., Ala. and w.

16. Q. velùtina Lam. Black O. (Pl. XIII, 5). Buds large, of about the dimensions of those of the Red O., but 5-angled or -sided, grayish-pubescent. Brts. apt to be downy or scurfy. Lvs. usually pubescent or scurfy below, with prominent tufts of hairs in axils of veins; with fewer lobes than in Red O., (7–9) usually 7, and lobes more oblong. Acorns have hemispherical cups and yellow meat. *Inner, living bark bright orange-*

yellow; hence sometimes called "Yellow O."; outer bark of old trunks *rough* and *black,* divided into polygonal chunks.   Best recognized in winter by its *stout, entirely grayish hairy, angled* buds; by these and by its yellow inner bark it can be distinguished from the Scarlet O., with which it may be confused.   Me. to Fla. and w.

17. **Q. ilicifòlia** Wang.   Bear or Scrub O. (Pl. XIII, 4).   A dwarf sp., usually shrubby, about 6 ft. in height, but sometimes tree-like, attaining 20 ft.; much branched and spreading; lvs. semipersistent, very variable in

Fig. 33.   *Quercus falcata,* Southern Red Oak.   Lf. × ½.

shape, tomentose beneath; brts. minutely downy; buds short, pointed, shining; acorns small, about ½ in. high, with usually saucer-shaped, sometimes deeper cups.   Apt to occur in barren soil and along the tops of ridges or peaks.   Me. to W.Va. upland to N.C. and w.

18. **Q. falcàta** Michx.   Southern Red O. (Fig. 33).   Resembles somewhat *Q. velutina* in the scurfy pubescence of brts. and lvs. but lvs. deeply cleft with lobes often falcate (i.e. scythe-shaped).   Acorn like that of *Q. velutina* but cup usually shallower.   Comparatively rare in s. N.J. and e. Pa. but common farther s.   Var. *pagodaefòlia* Ell. comes n. to s. N.J. and var. *trilòba* (Michx.) Nutt. with lvs. with only 3 lobes, is said to occur as far n. as L.I.

19. **Q. marilándica** Muenchh. Blackjack O. (Pl. XIII, 6). Best recognized by the very broadly obovate, almost triangular outline of its lvs.; buds about ¼ in. long, pointed, pubescent, red-brown. N.Y. to Fla. and w.

20. **Q. phéllos** L. Willow O. (Pl. XIII, 3). Buds sharp-pointed, about ⅛ in. long; acorns small, much like those of the Pin O., with shallow, saucer-shaped cups; narrow, entire, willow-like lvs. Occurs rarely on S.I. & L.I. and more commonly in N.J. s. to Fla. and w.[1]

The Willow Oaks form a distinct sub-group of the Black Oaks, with *entire* lvs. and acorns ripening in the second yr. Besides the above we have 21. **Q. imbricària** Michx., Shingle O., with wider, thicker lvs. than in the Willow O. and *shining* above. Del. and Pa. to S.C. and w. (A large tree in Brooklyn Bot. Gard.). 22. **Q. nìgra** L., Water O., with obovate lvs. often 3-lobed at apex or sometimes entire. Acorn with deeper cup than in Willow O. Del. and Pa. to Fla. and w., and 23. **Q. laurifòlia** Michx., Laurel O., with oblong or obovate lvs., dark green and shining above, light green beneath, glabrous, petioles yellow, acorn with saucer-shaped or sometimes deeper cup. Cape May, N.J. and s.

In our area the spp. most apt to occur are:

> White Oaks: *Q. alba*
> *Q. stellata*
> *Q. bicolor*
> *Q. prinus*

Q. alba is on the higher, drier sites, and *stellata* on sterile, sometimes rocky soil. *Bicolor* is to be looked for in swampy situations and *prinus* prefers rocky places.

> Black Oaks: *Q. rubra*
> *Q. palustris*
> *Q. coccinea*
> *Q. velutina*

Q. rubra is a common oak of higher rocky sites. *Coccinea* and *velutina* are not particular but are apt to occur in any locality except swampy soil. *Palustris* is common in swamps and wet ground but also may be found on dry sandy plains. *Q. prinoides* and *ilicifolia*, scrubby forms of the White and Black Oaks, respectively, occur in sterile soil or on tops of ridges. *Marilandica*, *phellos* and *imbricaria* are increasingly abundant toward the southern limit of our area. *Q. robur*, *petraea* and *cerris* are cultivated spp., not naturalized.

[1] It is sometimes difficult to persuade the layman that these trees *are* oaks since the lvs. resemble so closely those of a willow. But the facts that the buds are covered with *many* scales (willow has only one) and occur in a cluster at the ends of the brts. (as well as in lf. axils) and that the fr. is an acorn, rule out willows. As a matter of fact, in the northern part of its range, if the tree is an isolated specimen, acorns are often few or abortive, since the tree is usually self-sterile.

Oak Hybrids

Hybrids, that is, crosses of the above spp., in which two different spp. are the parents, sometimes occur.   They can be recognized by usually possessing characters of both parents, and also by the fact that the supposed parents grow somewhere in the neighborhood.   I have not included any of these: descriptions of the commoner ones will be found in the manuals.

As a matter of fact, hybrids of oaks are comparatively rare.   They are by no means as common as we might suppose if we consider the very long period of time that these trees or their forbears have been growing here.   It is quite probable that the pollen of any given sp. is more effective for fertilization on its own sp. than is the pollen of any other sp.   If this were not so we should have with us by now a motley array of every conceivable combination that would be almost impossible to untangle.

ULMÀCEAE — ELM FAMILY

Úlmus — Elm

Fr. surrounded by a membranous wing; lvs. usually inequilateral at base; buds with about 6 scales exposed, arranged in 2 ranks; fl. buds (in spring-fl. kinds) much larger than lf. buds; true terminal bud absent; stipule scars unequal; bundle scars 3, or in 3 groups.   (Pls. XV and XVI).

Key for Use during Season of Growth

Flowering in spring.
 Lvs. simply serrate or nearly so, nearly equal (not one-sided) at base, 1–3 in. long,
  elliptic-lanceolate; brts. gray; fl. buds spherical; bud scales tipped with long
  hairs; fr. small ($\frac{1}{2}$ in. long), glabrous (Pl. XV, 4) . . . . .6.   *U. pumila* 114
 Lvs. doubly serrate.
  Brts. glabrous at maturity; fr. small ($\frac{1}{4}$–$\frac{1}{2}$ in.) ciliate (Pl. XV, 1)
                           2.   *U. americana* 112
  Brts. pubescent; in no. 4 often glabrous in winter except at nodes.
   Buds black or very dark brown; cult.
    Lvs. large (3–6 in. long); bark in upper part of trunk and young brs. pale
     gray, smooth (Pl. XVI, 1) . . . . .4.   *U. glabra* 112
    Lvs. smaller (2–3 in. long); bark nearly black, rough, broken up into
     polygonal chunks (Pl. XVI, 2) . . . . .3.   *U. procera* 112
   Buds brown or red-brown; bud scales with darker margin.   Native.
    Fls. and fr. in elongated clusters (racemes); fr. ciliate; brts. often with
     corky wings (Pl. XV, 3) . . . . .5.   *U. thomasi* 114
    Fls. and fr. in dense clusters close to brt.; fr. not ciliate; brts. gray, without
     corky wings; inner bark mucilaginous, fragrant (Pl. XV, 2) . .1.   *U. rubra* 110
Flowering in autumn; lvs. simply serrate, small (1–2 in. long) (Fig. 34)
                              7.   *U. parvifolia* 114

1. **U. rùbra** Mühl. (*U. fulva* Michx.).   Slippery or Red Elm (Pl. XV, 2).   Lvs. *large* (4–8 in. long), very *rough* above to the touch, ovate-oblong, pointed, doubly serrate, pubescent below, not symmetrical at base; *brts. gray*, or sometimes brown, pubescent, and rough to the touch; *bark mucilaginous when chewed;* buds, especially the fl. buds, large, with rusty brown hairs, blunt; frs. $\frac{1}{4}$ to $\frac{3}{4}$ in. long, with wide wing, *not*

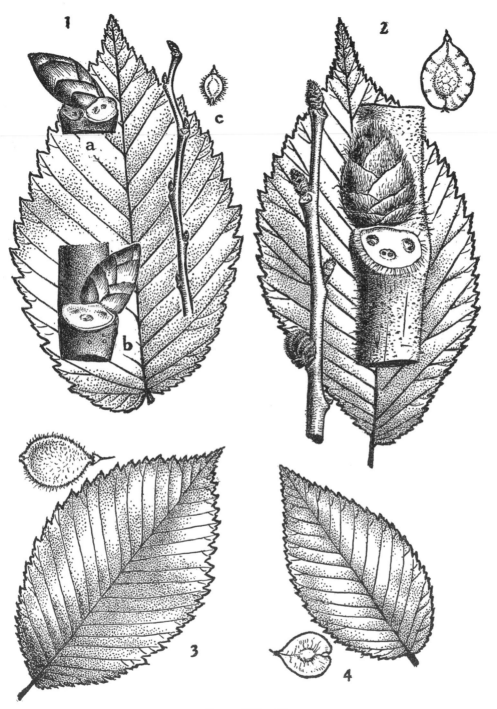

PLATE XV. Elms.

1. *Ulmus americana*, Amer. E. Lf. and brt. nat. size; a, uppermost lat. bud with remains of terminus of year's growth, below, at left × 5; b, lat. bud showing characteristic one-sided position above lf. scar × 5; c, fr. about nat. size.

2. *U. rubra*, Slippery E. Lf. and brt. nat. size; brt. and lat. bud × 5; fr. nat. size.

3. *U. thomasi*, Rock E. Lf. and fr. nat. size.

4. *U. pumila*, Siberian E. Lf. and fr. nat. size.

ciliate; borne in short-stalked, dense clusters.   Occasional in rocky woods. Que. to Fla. and w.

2. **U. americàna** L.   American E. (Pl. XV, 1).   Lvs. as in Slippery E. but smaller, not so rough to the touch, and smoother below; *brts. brown*, somewhat zig-zag, with elm taste, but *not mucilaginous;* buds red-brown, and fairly smooth, often situated a *little at one side of the lf. scar;* frs. much smaller than in last, less than ¼ in. long, *ciliate*, borne in *long-stalked clusters;* a larger tree than the last, with bark more deeply grooved; trunk typically dividing above into several leaders.   Common in moist soil, and much planted as a street tree.   Nfd. to Fla. and w.

Fig. 34.   *Ulmus parvifolia,* Chinese Elm.   Brt. with lvs. nat. size.

3. U. procèra Salisb.   English E. (Pl. XVI, 2).   Similar in a general way to the American E. but differs mainly as follows: lvs. smaller, and more pubescent below; buds darker or almost black; brts. generally downy; bark darker colored, and divided into small, irregular, polygonal plates; tree with typically a single main trunk, not dividing into a number of leaders.   Often cult. in several vars.

4. U. glàbra Huds.   Scotch or Wych E. (Pl. XVI, 1).   From Eu. and western Asia; has a *light gray* bark which stays smooth for a long time.   Occasionally some of the lvs. on vigorous shoots are 3-lobed at tip, or show a tendency to lobing.   Fr. comparatively large, about 1 in. long, not ciliate, with a wide wing.   Fl. buds almost spherical, the scales at apex tipped with long hairs.   Often planted as a street tree.

The Camperdown E., *U. glàbra* var. *camperdówni* Rehd., with pendulous brts. and round-topped head, is occasionally cult.   Many other cult. vars.

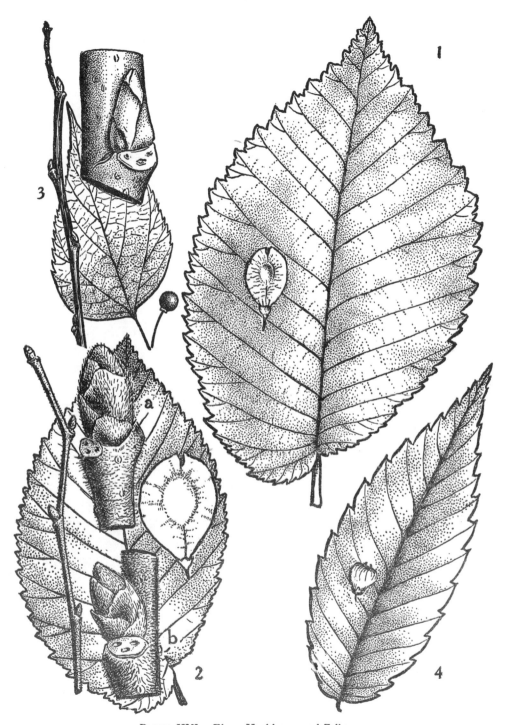

PLATE XVI. Elms, Hackberry and Zelkova.

1. *Ulmus glabra*, Scotch or Wych E.   Lf. nat. size; fr. × about ½.
2. *U. procera*, English E.   Lf. and brt. nat. size; fr. × 1½; a, uppermost lat. bud and brt. with stub at end of yr's growth below at rt. × 5; b, lat. bud and brt. × 5.   (Lf. is a little above average size.)
3. *Celtis occidentalis*, Common Hackberry.   Lf., fr. and brt. nat. size; lat. bud, lf. scar and part of brt. × 5.   Fr. a little smaller than average size.
4. *Zelkova serrata*, Japanese Zelkova.   Lf. and fr. nat. size.

Other spp. not so common are: 5. **U. thômasi** Sarg. Rock or Cork E. (Pl. XV, 3) with brts. often more or less corky; fls. (and frs.) in *racemes*, and pubescent fr., ¼–¾ in. long; Quebec and w. N.E. and w., occasional in w. N.Y. State on limestone ridges and ledges; fine specimen trees at Arnold Arboretum; 6. *U. pùmila* L. Siberian E. (Pl. XV, 4), now being much planted, notably in the shelter-belts of the middle w. where it is commonly called Chinese E., with slender gray or gray-brown brts., spherical buds; lvs. usually nearly equal at base, almost simply serrate, i.e., the teeth entire or with only one minute tooth; and 7. *U. parvifòlia* Jacq. Chinese E. (Fig. 34) with *simply serrate* small lvs., 1–2 in. long. Blooms in Aug. and Sept., the only elm in our area to bloom in fall. Good specimen in Prospect Pk., Brooklyn, near Willink entrance.

Zélkova serràta (Thunb.) Mak. Japanese Zelkova, (Pl. XVI, 4) a near relative of the E., is hardy and is rarely cult. in our area. Characterized by sharply and simply serrate lvs. with long-pointed teeth, and by the short stalked, *drupaceous* fr. Good living specimen at Brooklyn Bot. Gard. and one in Flushing, N.Y.

CÉLTIS — HACKBERRY

C. occidentàlis L. Common Hackberry (Pl. XVI, 3). Buds small, sharp-pointed, pressed close to brt.; true terminal bud lacking; stipule scars narrow; pith of brts. *closely chambered*, sometimes only at the nodes; *bark warty* (caused by local growths of cork); insect galls usually plentiful on lvs. and twigs, often causing "witches' brooms" on latter; lvs. ovate, long-pointed, serrate, unequal at base, with 3 prominent veins starting from base; fr. a drupe the size of a pea, with thin, sweet, edible flesh, yellow when ripe; a small tree. Que. to Fla. and w.

MORÀCEAE — MULBERRY FAMILY

All members of the family have milky sap.

MACLÙRA — OSAGE-ORANGE

M. pomífera (Raf.) Schneid. Osage-orange (Pl. XVII, 4). Medium-sized tree with yellow-brown bark; shining, entire, pointed lvs.; *axillary*, *simple* thorns; milky sap; and large multiple frs. the size of an orange, mostly falling green; buds small, globular; true terminal bud lacking; stipule scars small. Cult. and sparingly natzd. in our area. Used for hedges. Ark. to Okla. and Tex.

CUDRANIA

Cudrania tricuspidata (Carr.) Bur. is a shrub or small tree, to 50 ft., from China, Korea, and Japan, with ovate to obovate entire lvs., 1–3 in. long, sometimes 3-lobed at apex, axillary thorns, and small, red, spherical, edible fr. about 1 in. diam. Doubtfully hardy n. Related to Osage-orange and has been crossed with it (*Macludrania*).

BROUSSONÈTIA — PAPER-MULBERRY

B. papyrífera (L.) Vent. Paper-mulberry (Pl. XVII, 3). A medium-sized tree with smooth bark, gray with pinkish tinge, and milky sap; lvs.

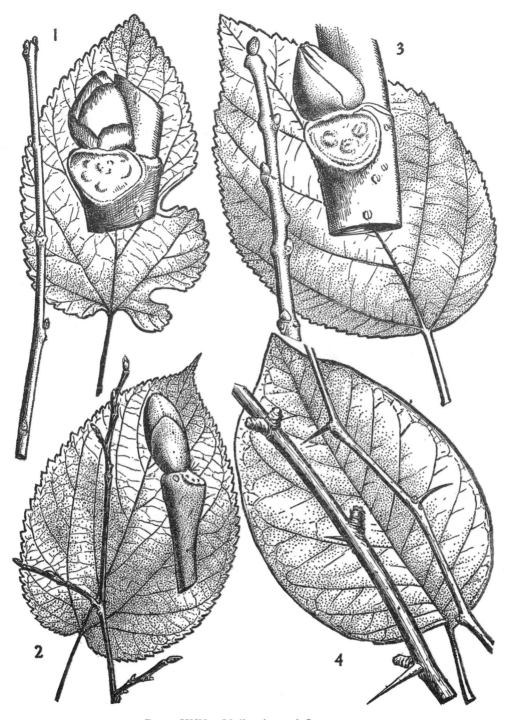

PLATE XVII.　Mulberries and Osage-orange.

1. *Morus alba*, White Mulberry.　Lf. and brt. nat. size; lat. bud and lf. scar × 5;
2. *M. rubra*, Red M.　Lf. and brt. nat. size; uppermost lat. bud, lf. scar and stipule scar × 5.
3. *Broussonetia papyrifera*, Paper-m.　Lf. and brt. nat. size; lat. bud, lf. scar and part of brt. × 5.
4. *Maclura pomifera*, Osage-orange.　Lf. and brt. nat. size; brt. showing thorns and **short growths terminated by buds** × 2.

often opposite, pubescent below, rough above, and often lobed; lf. scars rounded, elevated; buds covered by 2 scales, one striate; terminal bud lacking; pith white, with a thin, green partition at each node. China and Japan. Cult. and sparingly natzd., N.Y. to Fla. and w.

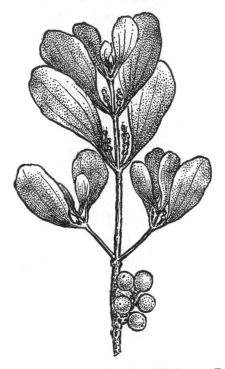

Fig. 35. *Phoradendron flavescens*, American Mistletoe. Fr. brt. nat. size.

Mòrus — Mulberry

1. **M. rùbra** L. Red Mulberry (Pl. XVII, 2). Lvs. broad, occasionally lobed, rough above, downy beneath; bark darker than in next; buds larger, with *green* tinge. Ripe fr. dark purple or red, pleasant. Native sp. in woods, but rare. Sw. Vt. to Fla. and w.

2. **M. álba** L. White M. (Pl. XVII, 1). *Lvs. smooth and shining*, often variously lobed; bark with yellowish tinge; buds red-brown, triangular in outline. Fr. white, pink, or purplish, insipid. Natzd. from Asia. Several vars. cult. Milky sap does not show clearly in winter.

3. *M. nìgra* L. Black M. is very similar to M. alba, but the lvs. are more apt to be undivided, deeply cordate at base, and scabrous above. Fr. *dark red.* W. Asia.

Fìcus càrica. L. Common Fig, with large lvs. 3–5 lobed, a native of w. Asia, is hardy in the southern part of our area and occasionally cult.

LORANTHÀCEAE — MISTLETOE FAMILY

⁰Phoradéndron flavéscens (Pursh) Nutt. Amer.-mistletoe (Fig. 35), is a much branched shrub half parasitic on various deciduous trees, notably oaks, i.e., has its own green lvs. but gets its sap from the tree on which it grows. Lvs. opposite, oblong or obovate, about 1 in. long, thick and evergreen. Fr. a small white drupe. N.J. and e. Pa. to Fla. and w. The true mistletoe (*Viscum album* L.) is similar, a native of Eu. and n. Asia.

The Dwarf-m., ⁰Arceuthòbium pusíllum Peck, is a small shrub not more than 1 in. in height, parasitic on the brts. of Spruce and sometimes Larch and White Pine, with greenish or brownish stems springing from the inner living bark of the host, and scale-like, opposite lvs. Often causes "witches' brooms" on conifers. Nfd. to n. N.J. and n. Pa. and w.

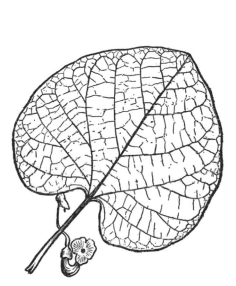

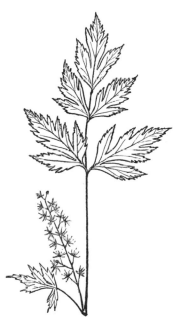

FIG. 36. *Aristolochia durior*, Dutch-man's-Pipe. Fl. and lf. $\times$ ½.

FIG. 37. *Xanthorhiza simplicissima*, Shrub-Yellowroot. Lvs. and fls. $\times$ ½.

ARISTOLOCHIÀCEAE — BIRTHWORT FAMILY

⁰⁰Aristolòchia dùrior Hill. Common Dutchman's-pipe. (Fig. 36). A tall, twining vine with very large (4–16 in. in diam.) kidney-shaped lvs.; the fls. curved like a Dutch-man's pipe. Buds superposed, hidden by lf. base. Pa. to Ga. and w. Often cult.; makes an excellent screen. Sometimes escaped n. of its range.

POLYGONÀCEAE — BUCKWHEAT FAMILY

The Fleecevine, *Polýgonum aubérti* (L.) Henry, with ovate or oblong-ovate lvs. lobed at base as in an arrowhead, and with wavy margins, abundant white or greenish-white fls. and fr. in late summer and early fall, climbs on supports by its twining stems and is sometimes cult.   Pith with a single diaphragm at nodes.   China.

RANUNCULÀCEAE — BUTTERCUP FAMILY

Most of the family are herbaceous, the typical lf. form being dissected, often after the plan of Buttercup lvs.

Three genera are woody or have woody representatives.

⁰ *Paeònia.*   Peony.   One sp., *P. suffruticòsa* Andr., Tree P., is commonly cult.   Low shrub attaining 6 ft., with bipinnate lvs. and large (4–12 in. in diam.) fls., white, rose, or red.   Nw. China.   Most of the Peonies, of which, in 1942, there were well over 3000 cult. vars., are herbaceous.

FIG. 38.   *Akebia quinata*, Fiveleaf Akebia.   Lf. and twig × ½.

⁰**Xanthorhìza simplicíssima** Marsh.   Shrub-Yellowroot.   (Fig. 37).   Low shrub, with yellow bark and root, sometimes planted for border of shrubberies.   N.Y. to Fla. and w.

⁰⁰**Clématis.**   Handsome, ornamental slightly woody vines.   About 150 spp. and vars. are cult.   The following are native in this area.

⁰⁰1. **C. virginiàna** L.   Virgin's Bower.   Fls. dull white.   Climbing by means of twining lf. stalks.   Lvs. 3-foliolate, *opposite*, stems finely pubescent.   Also known as Devil's-Hair, from the clusters of hairy frs. with threadlike extensions, the styles, clothed with long hairs.

2. ⁰⁰**C. verticillaris** DC.   Rock C. with handsome purplish fls., 6-sided glabrous brown stems and 3-foliolate, opposite lvs. occurs sparingly on rocky sides of talus slopes.   Que. to Va. and w.

3. ⁰**C. ochroleuca** Ait. Curly-heads, an erect sp. with simple or merely cleft lvs. and yellowish to purplish fls. occurs from se. N.Y. to e. Pa. and s.

Many other cult. spp. and vars. in which lvs. are usually compound; stems slightly woody at base.

LARDIZABALÀCEAE — LARDIZABALA FAMILY

AKÈBIA

⁰⁰*A.* **quinata** (Houtt.) Decne.   Fiveleaf Akebia.   (Fig. 38).   Graceful woody vine with dark colored fls. in spring, easily distinguished by the lvs., which have 5 emarginate,

entire lfts. all arising from a single point. The fr., rarely seen, resembles somewhat a small violet-colored banana 2–3 in. long. Sometimes cult. and occasionally natzd. C. China to Japan and Korea.

BERBERIDÀCEAE — BARBERRY FAMILY
BÉRBERIS — BARBERRY

Shrubs with red berries and a thorn (modified lf.) at each node. (Fig. 39).

The 2 spp. of barberry commonly seen are the Common B., [0]**B. vulgàris** L., a native of Eu. but natzd. in woods and fields, with berries in a long cluster, and with usually 3-pronged thorns; and the Japanese B., [0]*B. thunbérgi* DC., much cult. for hedges, with berries borne singly or in umbel-like clusters, and with usually simple thorns. Several vars. of the latter, which is becoming natzd., are cult.

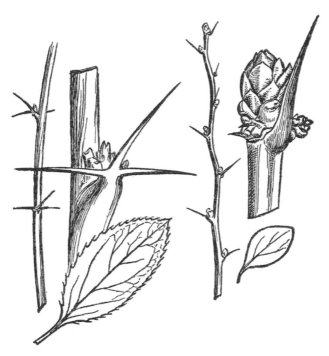

Fig. 39. *Berberis;* left, *B. vulgaris,* Common Barberry; rt. *B. thunbergi,* Japanese Barberry.

MAHÒNIA

The Oregon-grape, [0*]*M. aquifòlium* Nutt., a low shrub with evergreen, pinnate, spiny-margined lvs., and blue-black, glaucous frs., is occasionally cult.

MENISPERMÀCEAE — MOONSEED FAMILY
MENISPÉRMUM — MOONSEED

The Moonseed, [00]**M. canadénse** L. (Fig. 40), is a woody, dioecious climber with broad, peltate, lobed or angular, alternate lvs.; and small, black, fleshy frs., each with one crescent-shaped seed (hence the name); brts. more or less minutely fluted, slender; lf. scars elliptical or circular, raised, and with concave surface; buds very small, partly buried in bark. Occasional in woods and thickets in rich soil. Que. to Ga. and w.

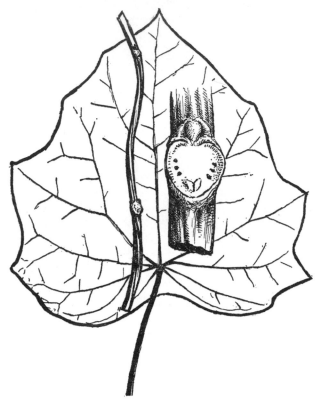

FIG. 40. *Menispermum canadense*, Moonseed. Lf. (often not so deeply lobed as this) and part of brt. nat. size. Bud and lf. scar × 4.

CERCIDIPHYLLÀCEAE — KATSURA-TREE FAMILY

The Katsura-tree, *Cercidiphýllum japónicum* Sieb. and Zucc. (Fig. 41), is becoming increasingly popular in cult. Recognized by its opposite or subopposite lvs. shaped somewhat like those of *Cércis*, the Redbud, hence the name, and by the red buds with only one of the scales exposed, and closely appressed to the stem. Japan. A beautiful small tree usually not more than 30 ft. tall; good specimens at N.Y. and Brooklyn Bot. Gardens; also at Kissena Pk., Flushing, N.Y.

MAGNOLIÀCEAE — MAGNOLIA FAMILY

MAGNÒLIA

Buds with only one scale showing, the inner scales (in lf. buds) alternating with the rudimentary lvs.; large true terminal bud; stipule scars linear, encircling brt. (Pl. XVIII); pith continuous, but sometimes with firmer diaphragms; brts. with aromatic odor.

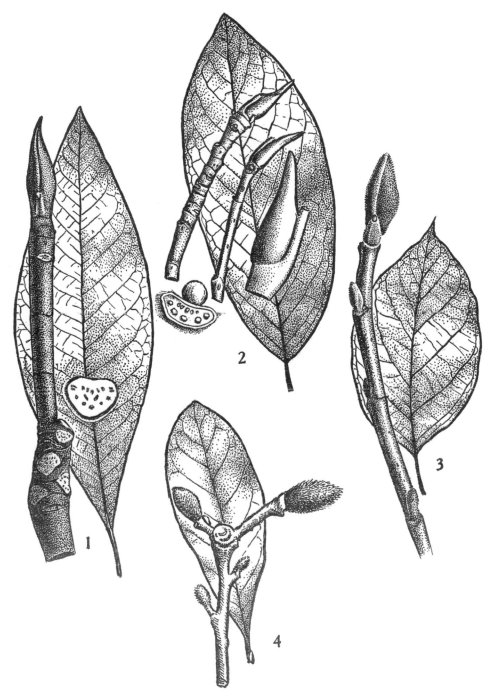

PLATE XVIII. Magnolias.

1. *Magnolia tripetala*, Umbrella M. Brt. nat. size; lf. × ⅛; lf. scar × 2.
2. *M. virginiana*, Sweet-bay. Lf. and brts. nat. size; lf. scar and lat. bud × 5; terminal bud × 3.
3. *M. acuminata*, Cucumber-tree. Brt. nat. size; lf. × ⅛.
4. *M. stellata*, Star M. Brt. and lf. nat. size.

Most of the cult. magnolias in our area are Asiatic, and bloom in early spring before the lvs. appear.   Perhaps the commonest is the Saucer M. (*M. soulangeàna* Soul.) (Fig. 42), supposed to be a hybrid form of two Asiatic Magnolias, with large pink fls. which fill the whole tree, a glorious sight; at about the same time the Yulan M. (*M. denudàta* Desrouss.) blooms with equally abundant large fls. of creamy white.   Somewhat earlier, even in March, the Star M. (*M. stellàta* Sieb. and Zucc.) (Pl. XVIII, 4), a shrubby sp., unfolds its masses of white or rose-tinged fls.   *M. kòbus*, DC. with white fls. and petals with a faint purple line outside, is not so commonly planted.   *M. liliflora*

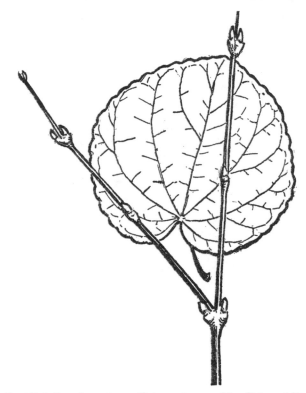

FIG. 41.   *Cercidiphyllum japonicum*, Katsura-tree, p. 97.   Brt. and lf. nat. size.

Desrouss., the Lily M., has purple fls. which often come out with the lvs.   Finally, the Oyama M., *M. sieboldi* (*parviflòra*) K. Koch, is an exception to these early blooming Asiatics; its Japanesy fls., with bright crimson stamens and pistils, appear much after the comparatively small, elliptic acuminate lvs., beginning in June.

The spp. native in the U.S., of which there are eight or nine, bloom after or with the appearance of the lvs.; and the fls. are therefore not so conspicuous, although often large.

Key to Principal Native Species

Lvs. evergreen; native from N.C. to Fla. and Tex. (See also no. 4)..5. *M. grandiflora* 123
Lvs. deciduous.
    Buds glaucous, long and tapering (lvs. 10–24 in.) (Pl. XVIII, 1)..3. *M. tripetala* 123

Buds more or less pubescent.
Lvs. very large (1–3 ft. long), oblong obovate, rare tree in n..2. *M. macrophylla* 123
Lvs. not so large.
Lvs. glaucous beneath, 3–5 in. long....................4. *M. virginiana* 123
Lvs. soft pubescent beneath, 5–10 in. long...............1. *M. acuminata* 123

1. **M. acumináta** L. Cucumber-tree (Pl. XVIII, 3). Large cone-shaped forest tree, much planted in parks and estates; lvs. abruptly short-inted, 5–10 in. long; buds pubescent; fr. knobby, 2–3 in. long, and narrow, resembling cucumber; fls. in May or June, fairly large but inconspicuous because somewhat similar in color to young foliage. W. N.Y. to Ga. and Ala. and w.

2. *M. macrophýlla Michx. Bigleaf M. A rare tree in n. but sometimes cult. Hardy to e. Mass. Recognized by enormous lvs., sometimes nearly 3 ft. long, and by tomentose buds; fls. white, nearly 1 ft. in diam., in May or June. W. Va. to Fla. and w.

3. **M. tripétala** L. Umbrella Magnolia (Pl. XVIII, 1). Characterized especially by glabrous, and usually glaucous, long-pointed buds, rather large lvs. (10–24 in.) clustered at tips of brts. Small tree with rather irregular, wide-spreading brs. Much cult. Pa. to Ga. and w.

4. **M. virginiàna** L. Sweet-bay (Pl. XVIII, 2). Small tree; lvs. half-evergreen, *glaucous* beneath; buds sparsely pubescent or silky; brts. bright green; fls. creamy white, in summer, very fragrant (Gardenia odor). Swamps, Mass. to Fla. and w., mostly near coast. Cult.

Fig. 42. *Magnolia soulangeana*, Saucer Magnolia. Lf. and bud × ½.

5. The Southern M., *M. *grandiflòra*, sometimes called the Evergreen M. or Bull-bay, is a magnificent evergreen tree much planted in the Southern States, but scarcely hardy n. of Del. A large specimen is growing in Brooklyn, N.Y. in a protected location.

LIRIODÉNDRON — YELLOW-POPLAR

L. tulipífera L. Yellow-poplar (Fig. 43). Also known as Tulip-poplar, Tulip-tree, and Whitewood. A tall, straight, forest tree, valuable for its timber; lvs. squarish at apex; pith white, diaphragmed; bud shaped like a duck's bill, smooth, covered by two valvate scales; true terminal bud present; stipule scars encircling brt.; frs. cone-shaped, conspicuous, and long persistent, often throughout the winter.   Deep, rich soil.   Mass. to Wis. and s.

FIG. 43.   *Liriodendron tulipifera*, Yellow-poplar, Tulip-tree.   Lf. and brt. × ½; end bud and brt. tip × 4.

CALYCANTHÀCEAE — CALYCANTHUS FAMILY

*⁰ *Calycánthus flòridus* L.   Sweet-shrub, also known as Carolina-allspice and Strawberry-shrub (Fig. 44) is unmistakable when in flower, from its dark red-brown, fragrant fls.   Lvs. simple and opposite, pubescent beneath.   Commonly cult.   Va. to Fla. and w.
⁰C. fértilis Walt. is similar but lvs. are *glabrous* and fls. are greenish-purple to red-brown. Pa. to Ga. and w.   Fr. peculiar; large and bag-like.

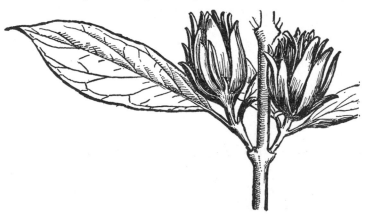

FIG. 44.   *Calycanthus floridus*, Sweet-shrub, Carolina-allspice, nat. size.

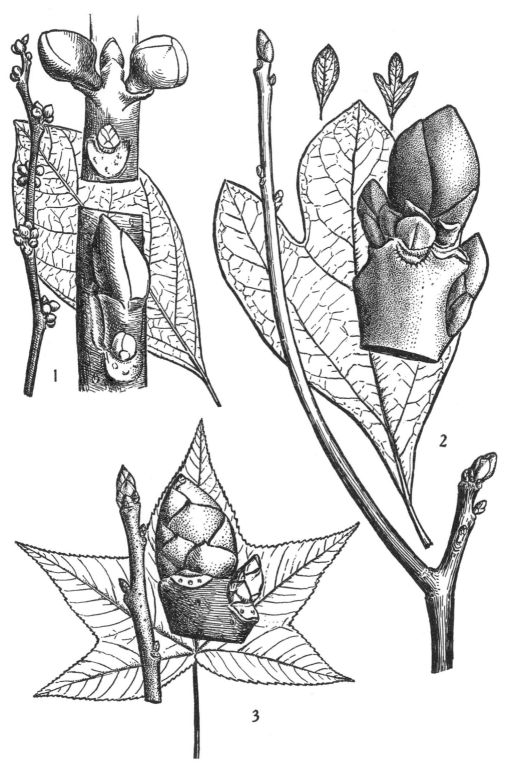

PLATE XIX. Laurel Family and Liquidambar. 1. *Lindera benzoin*, Spicebush. Brt. and lf. nat. size; parts of brt. with lf. and fl. buds and lf. scars × 10. 2. *Sassafras albidum*, Sassafras. Brt. and lf. nat. size; terminal bud and accessories × 5; top, two other forms of lvs. much reduced. 3. *Liquidambar styraciflua*, Sweet-gum. Brt. and lf. nat. size; terminal bud and accessories × 4.

ANNONÀCEAE — CUSTARD-APPLE FAMILY

Asímina trilòba (L.) Dun. Common Pawpaw. Small tree with simple, large, alternate, entire lvs. (6–12 in. long), distinguished especially by its true terminal naked winter bud, resembling somewhat that of the Witch-Hazel (Pl. XI, 3), flat and knifelike in outline, reddish tomentose; lateral buds superposed, globose and stalked, if fl. buds; or oblong and nearly sessile, if lf. buds; pith diaphragmed, sometimes indistinctly, or in age, chambered. Fls. with 3 sepals and 6 petals, and fr. edible, somewhat like a small banana. A good specimen at the Brooklyn Bot. Gard. W. N.Y. and Ont. to Fla. and w.

LAURÀCEAE — LAUREL FAMILY

SÁSSAFRAS — SASSAFRAS

S. álbidum (Nutt.) Nees. Common Sassafras (Pl. XIX, 2). Brts. and buds green and glabrous, with characteristic odor and taste; internodes conspicuously variable in length; true terminal bud present, large, with about 4 scales exposed; stipule scars lacking; lvs. simple, variously lobed or entire; old bark brown and fragrant, in old trees deeply furrowed into broad, flat ridges, which are sometimes cut by transverse lines; often shrubby, but sometimes a large tree, the trunk attaining nearly 6 ft. in diam. Sw. Me. to Va. and w. and, in var. *mólle*, (Raf.) Fern. with pubescent brts. and lvs. pubescent beneath, s. to Fla. and w.

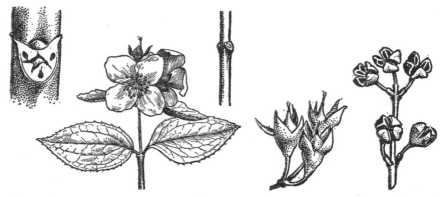

FIG. 45. *Philadelphus grandiflorus*, Mock-orange, "Syringa." Left, bud, mostly hidden beneath lf. scar, × 5; center, fl. brt. × about ½; right, brt. nat. size.

FIG. 46. *Philadelphus*, Mock-orange. Left, fr. brt. soon after fall of petals; right, winter appearance after dehiscence of fr.

LÍNDERA — SPICEBUSH

⁰**L. benzòin** (L.) Bl. Spicebush (Pl. XIX, 1). Brts. with characteristic spicy odor; buds of two kinds; fl. buds larger and globular, lf. buds smaller and pointed, with 2–3 scales showing; on brts. where fl. buds occur, they are typically arranged in pairs at the nodes, one on each side of a lf. bud; often extra or accessory fl. buds occur, as in Pl. XIX, 1, brt. at

left; true terminal bud and stipule scars lacking; lvs. simple, obovate. Me. to Fla. and w.

SAXIFRAGÀCEAE — SAXIFRAGE FAMILY

Contains 3 of our most popular ornamental shrubs and a fourth genus, *Ribes,* comprising the currants and gooseberries.

PHILADÉLPHUS — MOCK-ORANGE (Figs. 45, 46)

Fls. with 4 large petals in early June, usually very fragrant. Lobes of the sepals surmounting the ovary are persistent for some time after flowering (Fig. 46). Lvs. opposite, remotely toothed; brts. with fine vertical lines or ridges—in most spp. sloughing off the outer bark; pith solid. Much cult. Many spp., vars. and hybrids, one of the best being *P. coronàrius* L. *P. lemoìnei* is a popular hybrid. Unfortunately *Philadelphus* is also known popularly as "Syringa," which name should be avoided, since that is the botanical name for the genus of lilacs (p. 221). Eu., Asia, Amer.; several in s. and w. U.S. See under *Deutzia* for distinction between this and that genus.

FIG. 47. *Deutzia gracilis,* Slender Deutzia. Fr. brt. nat. size.

DEÙTZIA — DEUTZIA (Fig. 47)

Deutzia is usually somewhat smaller than the last, but with opposite lvs. which in some spp. are rough to the touch (on account of fine, appressed, stellate hairs), exfoliating bark and small clustered fls., the hemispherical, tardily dehiscent ovary at first resembling somewhat that of *Philadelphus. D. grácilis* Sieb. and Zucc., the Slender D. from Japan, with white fls., is commonly planted. Spp. from Asia and Mex.

During the winter months the clusters of dry frs. of *Deutzia* and *Philadelphus*, the former slow in splitting open, are a good means of identifying these shrubs, besides the fact that their brts. are opposite. In *Philadelphus* the frs. are *top-shaped*, the calyx lobes (Fig. 46) having withered away and mostly fallen off; in *Deutzia* the frs. are cup-shaped or almost globular and often the long styles are still conspicuous. Further, the brts. of *Deutzia* are usually hollow, while those of *Philadelphus* have a solid white pith.

HYDRÀNGEA — HYDRANGEA

Erect or sometimes climbing shrubs with exfoliating bark; lvs. opposite or whorled, rarely lobed; fls. in terminal clusters (corymbs or panicles) often with enlarged sterile marginal fls.; pith large, pale, continuous; buds globose-conical to oblong, with 2–3 pairs of scales showing. Much planted spp. are *H. paniculàta* and var. *grandiflòra*, the Panicled and "Peegee" H., these being easily recognized when in flower by the pyramidal flower clusters. In other spp. the fl. clusters are flat-topped. The genus is both Amer. and Asiatic. ⁰H. arboréscens L., with lvs. 3–8 in. long, occurs from s. N.Y. to Ga. and w. in rich woods and limestone soil.

The Climbing Hydrangea, *H. petiolàris* Sieb. and Zucc., with aerial rootlets, brown, flaky bark on older stems, fls. in clusters, some of them sterile and enlarged, and broad ovate acuminate lvs., is sometimes cult. Asia.

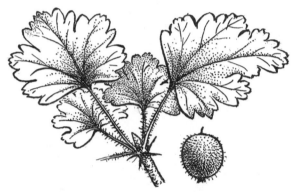

FIG. 48.   *Ribes grossularia*, Eu. Gooseberry.   One of several forms.   Brts., lvs. and fr. nat. size.

RÌBES — CURRANT.   GOOSEBERRY (Fig. 48)

The Gooseberries have prickly frs., and prickles or thorns on the stem, while the currants are without these appendages.   Lvs. alternate, simple, palmately lobed; fls. with parts in 5's; fr. a small, sour, juicy, many-seeded berry.

The common Red Currant of the gardens, with fr. in long drooping clusters, is *R. satìvum* Syme, a native of Eu.   The Eu. Black C., *R. nìgrum* L. and the Amer. Black C., ⁰R. americànum Mill. differ especially in the longer fr. clusters of the latter.   The Golden C., *R. aùreum* Pursh., a native of the far west, with yellow, spicy-fragrant fls., appearing with the lvs., is sometimes cult. as an ornamental shrub.   In northern woods the Skunk C., ⁰R. glandulòsum Weber, is frequent—a low, spreading shrub with characteristic skunk-like odor.   One of the commonest of the gooseberry section of the genus is ⁰R. cynósbati L., with slender, short, simple or 3-parted thorns, deeply 3–5 lobed lvs., and prickly, wine-colored, edible frs.   The Eu. Gooseberry, *R. grossulària* L., (Fig. 48) with mostly pubescent or bristly fr. is often cult. Eu. Many spp. of *Ribes* are carriers of the White Pine Blister Rust, especially *R. nìgrum* and most Amer. C. It is safer not to plant *any* currant or gooseberry where a 5-needled pine is growing.

HAMAMELIDÀCEAE — WITCH-HAZEL FAMILY

HAMAMÈLIS — WITCH-HAZEL

H. virginiàna L. Witch-hazel (Pl. XI, 3). Usually a shrub, but sometimes tree-like to 15 ft.; brts. zig-zag; buds stalked, the end bud crescent-shaped, tomentose, flattish; the lateral more cylindrical; lvs. oval, wavy-toothed, very one-sided at base; stipule scars unequal; fls. with long, narrow, crinkly, yellow petals, opening in Oct. and Nov.; fr. a 2-chambered capsule, shooting out its 2 seeds in the fall and remaining gaping open on the plant. Que. to Ga. and w.

The Chinese Witch-hazel, *Hamamèlis móllis* Oliv., deserves a wider acquaintance. A shrub, perfectly hardy in s. N.E., it blooms during mild periods in *Jan.* or even rarely in Dec., its bright yellow fls. resembling those of the common native sp. but of a lighter, more lemon-yellow hue. It gives a cheerful touch of color on drab wintry days. Later, in Feb. or Mar., a similar sp., the Japanese W., *H. japònica* Sieb. and Zucc., also quite hardy, is conspicuous with its masses of yellow fls. of a somewhat deeper hue. Both these shrubs are desirable members of any garden. *II. vernàlis* Sarg., the Vernal W., is another native sp. (Mo. to La. and Okla.) which blooms very early, sometimes even in Dec. with inconspicuous copper-colored fls. which have a spicy fragrance.

Parròtia pérsica C. A. Mev., a native of Iran, is a hardy, rare tree with lvs. and buds similar to those of the Witch-hazel. The bark of the trunk exfoliates, appearing somewhat like that of *Cornus kousa*, the Kousa Dogwood of Japan. Fls. with bright crimson stamens appear in Mar. Good specimens at the Brooklyn Bot. Gard. and Kissena Pk., Flushing, N.Y.

Corylópsis is a genus of hardy shrubs from Japan and China and desirable as ornamentals, with yellow fragrant fls. appearing before the lvs., which latter resemble somewhat those of *Corylus*, the Hazel, hence the name.

LIQUIDÁMBAR — SWEET-GUM

L. styracíflua L. Sweet-gum, Red-gum (Pl. XIX, 3). Large tree with *star-shaped lvs.;* buds pointed, shining, scaly, and red-brown or greenish; terminal bud large, $\frac{1}{4}$–$\frac{1}{2}$ in. long; bud scales with a minute point at tip, and fringed on margin; stipule scars lacking; brs. of second yr. and older often with corky ridges; brts. often brown above, green beneath; bark light gray and smooth on small trunks and brs., scaly and light brown on older trunks; fr. a spiny ball, often hanging on the tree through the winter. An important timber tree. Grows wild as far north as South Norwalk, Conn., and in N.Y. to the Hudson highlands, but is common in the southern States to Fla. and w. Often cult. and hardy much farther n. Autumnal coloration is beautiful.

FOTHERGILLA

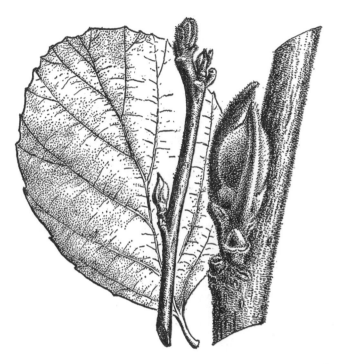

FIG. 48a. *Fothergilla major*, Large Fothergilla. Lf. and brt. nat. size. Part of brt. with fl. bud showing peculiar keeled scale enfolding fl., small bud below and stipule scar below that, all × 4.

The Large Fothergilla, *F. major* (Sims) Lodd., a shrub to about 9 ft., with lvs., buds, and fls. proclaiming its kinship to the Witch-hazels (Pl. XI, fig. 3, p. 95) is rarely cult. and deserves greater popularity. Conspicuous in very early spring with its masses of short spikes of white fls., mostly stamens (no petals) and in the fall with its orange to red lvs. Quite hardy in our area. Ga.

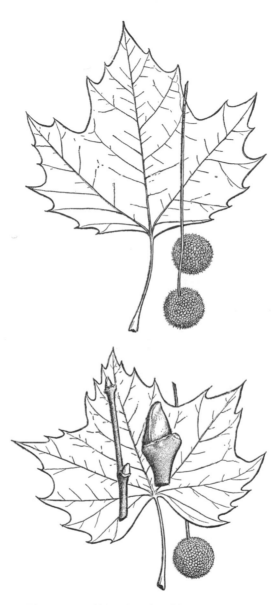

Fig. 49.   Above: *Platanus acerifolia*, London Plane-tree.   Lf. and fr. heads × ½.
(Lvs. are often less toothed and more distinctly 3-lobed than this.)   Below: *Platanus
occidentalis*, Sycamore.   Lf, fr. and brt. × ½.   Uppermost lat. bud showing lf. scar
surrounding bud, and, at rt., stub of end of season's growth, nat. size.

PLATANÀCEAE — PLANE-TREE FAMILY

PLÁTANUS — PLANE-TREE

Outer bark peeling off in thin plates of varying size, revealing whitish or yellowish inner bark; buds conical, covered by a single cap-like scale, and hidden under hollow base of petiole; true terminal bud lacking; stipule scars narrow, encircling brts.; lf. scars encircling bud; lvs. palmately lobed; fr. head a conspicuous ball, made up of many little hard frs., each surrounded by long hairs; large trees; the native sp. has the most massive trunk of all the deciduous trees in N.A. A diam. of 14 ft. has been recorded.

P. occidentàlis L. American Sycamore, Buttonball-tree (Fig. 49). Inner bark, where exposed, *white; lvs. shallowly 3–5-lobed* (lobes shorter than broad); fr. heads borne *singly* (very rarely in 2's), about 1 in. in diam., and not markedly bristly. Native in woods and fields, usually in moist soil. Easily recognized at a distance by its chalky-white patches of bark. Me. to Fla. and w.

P. orientàlis L. Oriental Plane-tree. Inner bark of a greenish white or grayish hue; *lvs. deeply 5–7- (rarely 3-) lobed* (lobes longer than broad); fr. heads 2–6, bristly, the smallest of those of the three spp., 1 in. or less in diam. Se. Eu. and w. Asia. Very rare in cult.

P. acerifòlia Willd. London Plane-tree (Fig. 49, p. 131). A hybrid between **P. occidentalis** and *P. orientalis*, sometimes resembling more the one, and sometimes more the other parent; inner bark of a *greenish* or *yellow* hue; lvs. 2–5 lobed, usually apparently 3-lobed, the middle lobe as long as, or slightly longer than broad; lobes not, or only sparingly toothed; lvs. have the general aspect of maple lvs. (hence the specific name); fr. heads usually in 2's (rarely 3's) bristly, about 1 in. in diam. The sp. commonly planted along streets; seems better adapted to city conditions than any other tree.

ROSÀCEAE — ROSE FAMILY

So many trees and shrubs of diverse appearance belong to the Rose Family that a word about their family connection is needed. The principal characters which unite all these in one family are that the stamens are usually *numerous* and borne *on the calyx*, (or, more exactly, on the edge of a ring lining the top of the calyx tube). Further, the lvs. are alternate, usually with conspicuous stipules. The fls. are perfect, i.e. both stamens and pistils are in the same fl., the petals are usually 5, and the calyx usually of 5 lobes or segments united at the base.

A large family, which may be divided into several tribes, as follows:

I. Spiraea tribe; with dry frs. P. 133.

II. Apple tribe; with characteristic fleshy fr. (pome); includes pears, quinces, shadbush, hawthorns, etc. P. 134.

III. Rose tribe; including *Rubus*, (blackberries and raspberries) and roses. P. 142.

IV. Prunus tribe; includes cherries, plums or prunes, peach, nectarine, and almond. P. 146.

I. SPIRAEA TRIBE

PHYSOCÁRPUS — NINEBARK

The Ninebark, [0]P. opulifòlius (L.) Maxim, (Fig. 50) is a natives hrub, often cult. Recognized by 3-lobed, serrate, alternate lvs.; bark exfoliating in thin strips (hence the name); small white fls. in umbel-like clusters in June, and reddish clusters of fr. pods in late summer. Que. to S.C. and w.

FIG. 50. *Physocarpus opulifolius.* Common Ninebark. Brt. with fls. nat. size; above, left and right, dehisced fr. cluster and single fr. enlarged; left side, winter brt., nat. size; below at right, lat. bud × 5.

FIG. 51. *Spiraea prunifolia,* var. *plena,* Double Bridal-wreath S. Fls. and brts. nat. size.

SPIRÀEA — SPIRAEA

Two spp. of Spiraea are common, native, low shrubs of our area: the Meadowsweet. [0]S. latifòlia (Ait.) Borkh., with lvs., frs., and purplish brts. glabrous, and white or pinkish fls., and the Hardhack, [0]S. tomentòsa L., with lvs., frs., and brts. rusty woolly (hence the specific name), and fls. mostly deep rose-color; true terminal bud lacking. P.E.I. to N.C., Ga. and w. Many other spp. are cult., some of the most popular being: ⁰S. *prunifòlia* Sieb. and Zucc., the Bridalwreath S., with white fls. early flowering, usually seen in the double form (Fig. 51); ⁰S. *vanhoúttei*, Vanhoutte S. with white fls., in mid-June, and peculiar small rhombic lvs.; and ⁰S. *japónica*, Japanese S., a lower shrub with usually pinkish fls.

SORBÀRIA — MOUNTAIN-ASH SPIRAEA

The Mountain-Ash S., ⁰S. *sorbifòlia* (L.) Br. A. (Ural Falsespirea SPN) a native of Siberia, has long been cult. and is often seen in and around old cemeteries in N.E. The pinnate lvs., somewhat like those of the Mountain-Ash, *Sorbus*, (hence the name) and the showy cluster of small white fls. in early June, are the chief characters. Travels fast underground, shooting up from the rootstocks.

EXOCHÓRDA — PEARLBUSH

The Pearlbushes, ⁰E. *giráldi* Hesse, and other spp., are handsome ornamental shrubs from Asia with rather large white fls. in terminal clusters; recognized by the characteristic lobed or strongly angled fr. pods (Fig. 52).

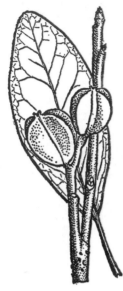

FIG. 52. *Exochorda Giraldi*, Redbud Pearlbush. Lf., brt., and fr. nat. size.

II. APPLE TRIBE

All genera have frs. of the structure of an apple; i. e., with a fleshy outer part, and inside this the carpels (cells of the ovary) which are papery, leathery, or bony, forming the so-called "core," with the remnants of the calyx lobes often showing at the tip of the fr. i. e. opposite the stem end. This kind of fr. is called a *pome*.

COTONEÁSTER — COTONEASTER

Named from *cotonea*, Latin for quince, and suffix *aster*, meaning "kind of," the lvs. of some spp. resembling those of the quince. Ornamental shrubs much planted for their unusual habit and attractive frs. The Rock Cotoneaster, ⁰*C. horizontàlis* Decne, (Fig. 53) a half-evergreen low shrub with horizontally spreading stems, admirably adapted for filling in spaces between rocks in rock gardens, is a commonly cult. sp. China. Many other cult. spp. See Rehder (49) 347–357 and Wyman (61) 136–140.

FIG. 53. *Cotoneaster horizontalis*, Rock Cotoneaster. Brt. with fr. nat. size.

PYRACÁNTHA — FIRETHORN

The Firethorn (Fig. 54) is characterized especially by its red or orange, small, apple-like fr. with persistent calyx, by the thorny brs. and *evergreen* lvs. The sp. most commonly planted is ⁰*P. coccínea* Roem. Scarlet Firethorn. Var. *lalándi* is handsome, being laden with bright orange fr. in the fall. Eurasia.

CRATAÈGUS — HAWTHORN

Small trees or shrubs, usually with thorns in the axils of the lvs., and small, apple-like frs.; buds rather small, rounded. Several native spp. grow in our area, but they are difficult to distinguish without fls. and frs. They can be identified as hawthorns by the axillary thorns (i. e. borne in the lf. axils) and by the frs.

The English H., *C. oxyacántha* L. (Fig. 55) has small, deeply lobed lvs., and mostly white or pink fls. — the "May" tree of English literature. Eu. N.Afr. Var. *pauli*, Rehd. with double, bright scarlet fls. is one of our most beautiful small trees. **C. phaenopỳrum** (L. f.) Med., the Washington Thorn, with scarlet fr., is another handsome sp. much planted. Native Pa. to Fla. and w. W. N.Y. State, near Route 20, a fertile field for study of various spp.

SÓRBUS — MOUNTAIN-ASH

Small trees, with clusters of attractive, bright red frs. in the fall, appearing like berries, but in reality like small apples in structure.

S. aucupària L.   European Mountain-ash (Rowan-tree).   (Fig. 56).
A small tree with *pinnate* lvs. resembling those of the ash, but *not oppo-site;* lfts. acute but not long pointed, pubescent beneath, at least when young; terminal buds much larger than lateral, pubescent with long, matted hairs; outer bud scales not gummy; bundle scars 3 or 5.   Often cult.   Eurasia.

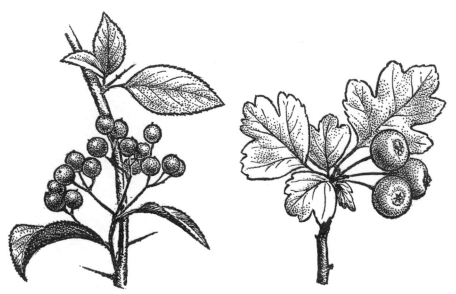

FIG. 54.   *Pyracantha coccinea*, var. *la-landi*, Laland Firethorn.   Fr. brt. nat. size.

FIG. 55.   *Crataegus*, Hawthorn, close to *C. oxyacantha*, English H.   Fr. brt. nat. size.

S. americàna Marsh.   Amer. M. has long pointed acuminate lfts., *smooth* beneath; outer bud scales glabrous and gummy; otherwise quite similar to the foregoing.   Nfd. to Ga. and w.

Sorbus decòra (Sarg.) Schneid. (*Pyrus decora* (Sarg.) Hyland), occurs in the northern part of our area, with S. americana, but with lfts. shorter and more rounded at tip and frs. about twice as large (⅓–½ in. in diam.)   More handsome and rounded in habit than S. americana.

S. hýbrida L., the Oakleaf M., a form in which the lvs. are tomentose on the under side, pinnate only toward the base, the upper part being lobed or sometimes the whole lf. only lobed, is occasionally cult.

ARÒNIA — CHOKEBERRY

Shrubs of wet places, occasional in drier soil; true terminal buds, with about 5 scales exposed; bundle scars 3.   The following as well as inter-mediate forms are common.   Aronia is sometimes classed as a subgenus of *Pyrus*, the pear.

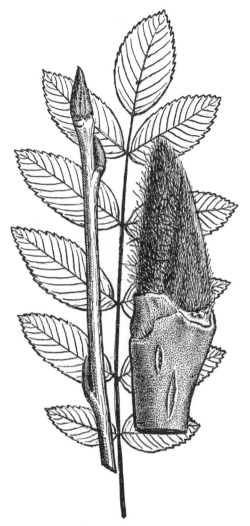

FIG. 56. *Sorbus aucuparia*, Eu. Mountain-Ash, Rowan. Brt. and lf. nat. size; bud × 2½.

⁰**A. arbutifòlia** (L.) Elliott. Red Chokeberry (Fig. 57). A shrub with clusters of small, apple-like frs.; buds carmine, sometimes with greenish tints, long, narrow, and sharp-pointed; lvs. elliptic, pubescent beneath, with many rounded teeth, and with tiny black glands along midrib on upper side (not shown in fig.); frs. bright red. N.S. to Fla. and w. ⁰**A. prunifòlia** (Marsh.) Rehd., Purple-fruit C., has fr. a little larger than that of the last and very dark red or purple. N.S. to Fla. and w.

⁰**A. melanocárpa** (Michx.) Elliott.   Black C.   Similar to last spp. but fr. is very dark purple or nearly black, and lvs. are *smooth beneath*.   Nfd. to S.C. and w.

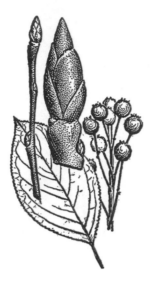

FIG. 57.   *Aronia arbutifolia*, Red Chokeberry.   Lf., brt., fr. nat. size; terminal bud × 4.

FIG. 58.   *Photinia villosa*, Oriental Photinia.   Flowering brt. nat. size; left, winter brt., nat. size; above, topmost lat. bud with stem end at rt.; below, lat. bud × 4.

PHOTÌNIA — PHOTINIA

The Oriental Photinia, *P. villòsa* (Thunb.) DC. (Fig. 58) is an ornamental small tree or shrub with white fls. in dense clusters (corymbs or short panicles), simple, finely and sharply serrate lvs. villous beneath, turning red in fall, and small apple-like fr., bright red, persistent. Japan, Korea, China.

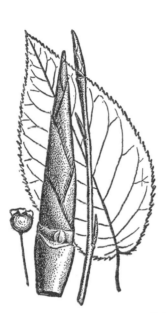 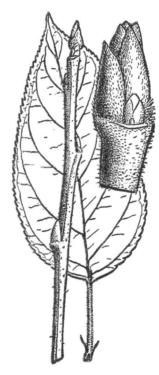

FIG. 59. *Amelanchier arborea* (Michx. f.) Fern. Serviceberry, Shadbush, Juneberry, lf., brt., and fr. nat. size; terminal bud × 4.

FIG. 60. *Malus pumila*, Common Apple. Brt. and lf. nat. size; bud × 4.

AMELÁNCHIER — SERVICEBERRY

In late Apr. or early May, a few days before the Flowering Dogwood blossoms, white patches in woods and swamps signalize the blooming of the Serviceberry, Shadbush, or Juneberry. Bark smooth, colored somewhat like that of Hornbeam (*Carpinus*); bud resembling that of Chokeberry (*Aronia*), slender and long pointed, but larger, usually greenish, often with a pinkish tinge; terminal bud much larger than lateral; bud scales sometimes twisted; bundle scars 3; stipule scars lacking.

A. arbòrea (Michx. f.) Fern. Serviceberry, Shadbush, Shadblow (Fig. 59) (= *A. canadensis* var. *Botryapium* of 7th Ed. of Gray). Buds ¼–½ in. long; lvs. small and densely white-woolly beneath at flowering time, heart-shaped or rounded at base; fr. reddish purple, insipid. N.B. to Fla. and w.

  A. canadénsis (L.) Medic. (*A. oblongifòlia* T. & G.) with oblong or oblong-obovate lvs. and blackish, sweetish fr. is also common. Me. to Ga. Several other spp. are native and common, and lvs. and fr. are needed for identification. See Gray's Manual of Botany, 8th ed. pp. 760–767, 1950.

Fɪɢ. 61. *Chaenomeles lagenaria*, Japanese Flowering-quince. Flowering brt. nat. size.

Màʟᴜs — Aᴘᴘʟᴇ

  M. pùmila Mill. Common Apple. (Fig. 60). Brts. more or less woolly at least toward tip; buds grayish, hairy at tip, blunt, terminal bud present, much larger than the lateral; stipule scars lacking; bundle scars **3**; lvs. oblong-ovate, pubescent beneath. The common "eating apple" of Eurasiatic origin, and now self-sown throughout the U.S. The Baldwins, Greenings, Pippins, etc., are cult. vars. of this sp. and can not be relied on to grow true to seed, but must be grafted. The wild apples coming from chance-sown seed, therefore, usually bear dwarfed, knubbly fr. Eu. and w. Asia.

  Many spp. of Malus are cult. for their fls. and ornamental fr., especially the "crabs," which have abundant fls. and small yellow or red fr. The Japanese Flowering Crab-apple, *M. floribúnda* Sieb., with carmine fl. buds changing to pale pink and then to nearly white, is one of the most popular ornamental shrubs or small trees, blooming early in May. The Carmine Crabapple, *M. atrosanguínea* (Spaeth) Schneid., a hybrid form, is even more handsome, with pinkish or rose-purple fls. — otherwise very similar. Many of these cult. ornamental apples are planted along Merritt and Hutchinson R. Pkwys. in sw. Conn. and Westchester Co., N.Y.

CHAENOMÈLES — FLOWERING-QUINCE

The Japanese Flowering-quince, *Chaenomèles lagenària* (Loisel). Koidz. (Fig. 61), has sharply serrate, simple lvs.; diffuse, often thorny, glabrous brts., scarlet-red varying to pink or white fls. and spherical to ovoid fr. which is yellow or yellow-green and fragrant. Sometimes used for preserves. Lvs. half-evergreen. Often called *C. japonica* in horticultural usage. China, cult. in Japan.

CYDÒNIA — QUINCE

The Quince, *C. oblónga* Mill., with entire, deciduous lvs. and white or pinkish fls. is often cult. for its large yellow-downy, fragrant frs. used in jelly and preserves.

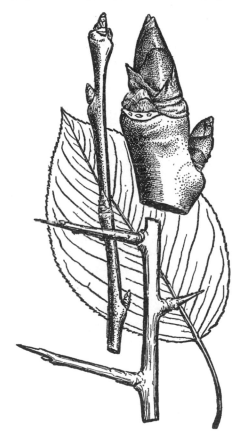

FIG. 62. *Pyrus communis*, Common Pear. Thornless and thorny brts. and lf. nat. size; terminal bud and accessory parts × 5.

PỲRUS — PEAR

P. commùnis L. Common Pear (Fig. 62). Often with thorn-like short brs.; buds usually glabrous, sharp-pointed; stipule scars lacking; bundle scars 3; lvs. rounded ovate to elliptic, *smooth beneath*. In the same category as *Malus pumila*, i. e., of Eu. and Asiatic origin, and long cult., but not so commonly found in a wild state.

III. ROSE TRIBE

Including *Rubus* (Blackberries and Raspberries) and Roses

May be divided into two groups: in (1) below, containing *Kerria*, *Rhodotypos*, and *Rubus*, the pistils are borne on a flat or convex receptacle; in (2) (p. 145), containing *Rosa*, the pistils are enclosed in a tubular or urn-shaped receptacle.

GROUP I

KÉRRIA — KERRIA

The Japanese Kerria, [0]*Kérria japónica* (L.) DC. (Fig. 63) is recognizable by its slender, zigzag, angled or ridged *green* stems; pith very wide and spongy; lvs. alternate, doubly serrate, with stipules, and fls. either single or double; also known as Japanese-Rose.   Commonly cult.   Grows to about 6 ft.   China and Japan.

RHODOTỲPOS — JETBEAD

The Jetbead, [0]*Rhodotỳpos scándens* (Thunb.) Mak. (Fig. 64), commonly cult., can be easily told in the winter from its 4 black, shining, small drupes, where the fls. were borne the preceding season.   Lvs. *opposite*, doubly serrate; fls. white, 4-petaled.   Grows to about 6 ft.   Japan and China.

FIG. 63.   *Kerria japonica*, Japanese Kerria.   Brts. with single and with double fls. nat. size.

RÙBUS — BLACKBERRY, RASPBERRY

In this genus the receptacle is convex, sometimes cone-shaped; basal portion of petiole persistent; buds commonly superposed, the lower one covered by the petiole base.   (Figs. 65, 66).

RASPBERRIES

The raspberries are either without prickles, or their prickles are usually weak and bristle-like. The fr., composed of many tiny frs. (*aggregate fr.*), is like a thimble in shape, and when picked, leaves the cone-shaped receptacle on the plant; lvs. usually compound.

FIG. 64. *Rhodotypos scandens*, Jetbead. Fl. brt. and fr. nat. size.

⁰R. idaèus L., var. strigòsus (Michx.) Maxim. Amer. Red Raspberry. Lfts. 3–5; coarsely doubly-serrate, white-tomentose beneath; young shoots densely bristly: bud glabrous and often glaucous. Fr. red, pleasant. Nfd. to w. Va. and w. In var. canadénsis Richards., the young shoots are covered with an ashy-gray down as well as being bristly. Lab. to N.E., to mts. of N.C., and w.

⁰R. occidentàlis L. Black or Blackcap R. (Fig. 65). Has glaucous, prickly stems, and lvs. white beneath. Que. to Ga. and w.

⁰R. odoràtus L. Purple-flowering R. Fragrant Thimbleberry (Fig. 66). A handsome sp. with large, purple fls., simple, large-lobed lvs., and light brown, loose, shreddy bark. It has no prickles, but its stems, when young, are covered with glandular hairs. Often cult. Que to Ga. and w.

The Wineberry, ⁰R. phoenicolàsius Maxim., a native of Japan, with stems and petioles covered with soft, reddish, glandular hairs, a few prickles also on the stems, is becoming more frequently natzd. in our area.

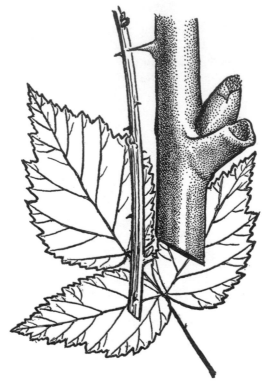

Fig. 65. *Rubus occidentalis*, Blackcap Raspberry.   Lf. and brt. nat. size; lat. bud, with persistent base of lf. stalk × 5.

Blackberries

The blackberries have (mostly) strong prickles and compound lvs. The fr., also aggregate, when picked, includes the pulpy receptacle. Many spp. are highly variable.

⁰R. híspidus L. Swamp Blackberry or Dewberry. Creeping, in swamps or moist woods, with prickles pointed backwards; lvs. smooth on both sides, somewhat leathery, usually evergreen. N.S. to Ga. and w.

⁰R. flagellàris Willd. Northern Dewberry. Creeping, on dry soil, with stronger prickles pointed backwards; lvs. thinner than in the last, and may be slightly soft-hairy below. Very variable. Que. to Va. and w. (= R. villosus of Gray, ed. 7).

Several spp. of high blackberries are common in our area. They are difficult to distinguish when not in fl. and fr. One of the commonest is:

⁰R. alleghaniénsis Porter. Allegany B. Distinguished particularly by erect, glandular pubescent stems, with stout prickles, by prickles and

glandular hairs on the fl. and fr. stalks, and by lfts. sparingly pubescent above, velvety beneath, green on both sides.   N.B. to upland N.C. and w.

Potentílla — Cinquefoil

Most spp. are herbaceous, but the Bush C., ^oP. fruticòsa L., a low shrub with shreddy bark, pinnate lvs. and attractive, yellow, rosaceous fls. is often seen in bloom in our area in pastures and uplands from early summer until Sept.

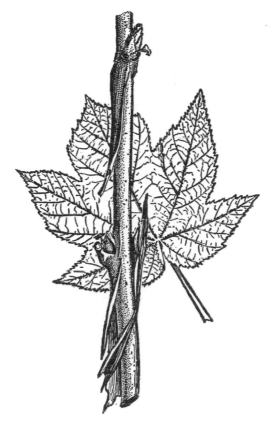

Fig. 66. *Rubus odoratus*, Purple-flowering Raspberry (Fragrant Thimbleberry SPN).   Brt. nat. size; lf. × ½.

GROUP II

Ròsa — Rose

Shrubs, with stems usually prickly, lvs. usually pinnately compound. Pistils enclosed in a tubular or urn-shaped receptacle which in fr. somewhat resembles a little apple externally ("rose-hip"), but is really a fleshy structure enclosing many little achenes.

When in flower, wild roses are easily recognized by their 5 large, pink, rose-colored, or crimson petals, (yellow in some exotic spp.) but without the fl. and fr. the spp. are difficult to identify.   Most of the wild spp. belong to the section *Carolinae*, in which the sepals after flowering are spreading, later deciduous, and achenes are borne only at the bottom of the receptacle.   Of these the following are common: ⁰**R. palústris** Marsh., Swamp R., usually with a pair of hooked prickles at each node, and 5–9 finely serrate lfts., the stipules rolled inwards.   N.S. to Fla. and w.   ⁰**R. virginiàna** Mill., Virginia R., with 5–9 more coarsely serrate and shining lfts., the stipules being flat and dilated. Nfd. to Ala. and w.   The Meadow R., ⁰**R. blánda** Ait., has no prickles or very few. Que. to Pa. and w.

Fig. 67.   *Rosa rugosa*, Rugosa Rose.   Fl. and lvs. × ½.

The Sweetbrier R., ⁰**R. eglantèria** L., is often natzd.   Eu.   Has strong, hooked, pinkish or salmon prickles; lvs. glandular on both sides, unmistakable from the apple-like odor when bruised or rubbed.   The Rugosa R., ⁰*R. rugosa* Thunb. (Fig. 67), a native of China, Korea, and Japan, is occasionally natzd.   Fls. large (2½–3½ in. in diam.) with lvs. very rugose above and very large fr. often more than 1 in. in diam.   The Scotch R., ⁰*R. spinosíssima* L., with a multitude of small lvs. with 5–11 lfts., densely prickly and bristly stems and white, pink, or yellow fls. is a common escape from cultivation.   Eu. and w. Asia.

Some of these roses and other exotic spp. are the parents of our cultivated roses — tea roses, hybrid perpetuals, climbers, ramblers etc. — over 5000 vars. having been named.

PRÙNUS TRIBE

Prùnus — Cherry, Plum, Peach, etc.

Trees or shrubs; lvs. simple, mostly toothed; fr. (a drupe) fleshy, with one stone inside; bark and lvs. of all spp. with a characteristic bitter flavor; buds usually with 4 or more scales exposed; terminal bud usually present, but lacking in the plums; bundle scars 3; stipule scars indistinct. In the cherries the stone is spherical, but in the plums, apricot (*P. armeníaca* L.) and peach, the stone is flattened.   The prune is a kind of plum that dries easily without decaying.

P. serótina Ehrh. Black Cherry (Pl. XX, 4). *Lvs.* simple, *thick, shining,* serrate with *incurved teeth* (see fig.); frs. black, in racemes; buds bright reddish brown, shiny, about 4 scales showing; brts. with bitter taste like that of cherry pits; bark on young trees reddish brown and smooth, on older brs. and young trunks marked with horizontally elongated lenticels, on old trunks covered with small, red-brown scales. One of the commonest trees in our area. Que. to Fla. and w.

P. virginiàna L. Choke Cherry (Pl. XX, 2). Usually a shrub, sometimes a tree; buds rather large (¼ in. long) sharp-pointed, paler than in last sp.; the margins of the scales light gray; taste of bark cherry-like, but different from that of Black C.; *lvs. thin,* sharply serrate, with *teeth pointing outward* (see fig.); frs. also in racemes, dark red, very puckery, ripening earlier than in *P. serotina.* Nfd. to upland of N.C. and w.

P. àvium L. Mazzard, Sweet C. (Pl. XX, 1). Lvs. usually with rounded, unequal teeth, pubescent on veins beneath; brts. stout, and of a paler hue than in last two; bark of young trees and brs. somewhat like that of Black C., but lenticels are larger and more yellowish; frs. in umbel-like clusters. Buds glossy, the fl. buds more ovoid. The sweet, dark red C. of the garden, native in Eu. and w. Asia, now natzd. over a large part of eastern U.S.

P. cérasus L. Sour C. is similar to last but has smaller lvs., smooth beneath, dull buds and sour, red fr. Natzd. from P.E.I. to Mich. and s.

P. pérsica (L.) Batsch. Peach. Lvs. smooth, long, narrow, tapering to a long point; brts. smooth, reddish on upper side, green beneath; buds pubescent; bark somewhat similar to that of cherries; fr. velvety. China. Often escaped.

P. pensylvánica L. f. Pin C., Wild Red C. (Pl. XX, 3) has long, narrow, long-pointed, finely serrate, smooth lvs. and small, translucent, red frs. in umbel-like clusters. More abundant northward and at higher elevations. Nfd. to n. Ga. and w.

The Sand or Dwarf C., ⁰**P. pùmila** L., a dwarf sp. usually less than 3 ft., with narrow oblanceolate lvs., occurs rarely in w. N.Y.

⁰**P. marítima** Marsh. Beach Plum (Pl. XX, 5). A low, spreading shrub with brts. velvety and buds slightly so; lvs. pubescent below, and red or purple frs. with a bloom, in umbel-like clusters, growing on and near sea beaches. Me. to Del.

A rather common wild plum in our area is **P. americàna** Marsh, Amer. P. a small tree with spreading brs., obovate, acuminate, sharply serrate lvs.; glabrous brts.; red-brown buds; white fls.; and red, rarely yellowish fr. W. N.E. to Fla. and w.

The "Flowering Almond," *Prunus glandulosa* Thunb. (Almond Cherry SPN) (Fig. 68) is a dwarf form (3–4 ft.) commonly planted as an ornamental, with early-blooming pink fls., the double form being most popular. The fls., about an inch in diam., are single or in 2's on short brts. along the main shoot, making a long wand-like display. Lvs. oblong-lanceolate, serrulate, mostly glabrous.

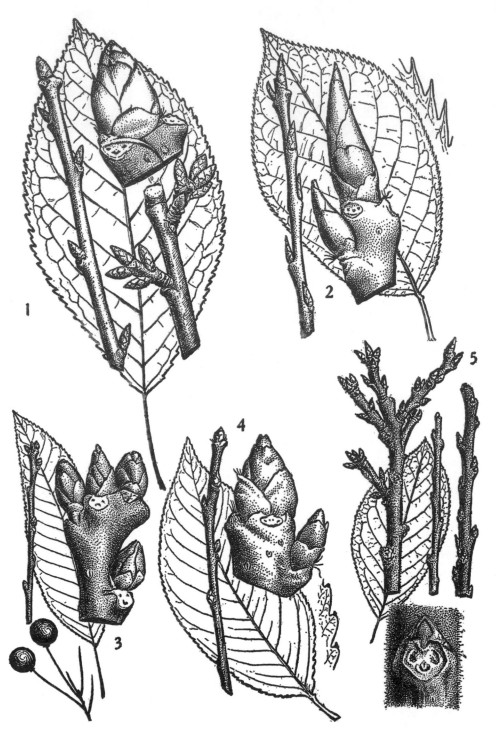

PLATE XX. Prunus: Cherries and Plum. 1. *Prunus avium*, Sweet Cherry, Mazzard. Lf. and brts.: left, vegetative brt.; rt., brt. with fl. buds nat. size; terminal bud and lf. scars × 5. 2. *P. virginiana*, Choke Cherry. Lf. and brt. nat. size; bud and accessory parts × 4; upper rt., lf. teeth enlarged. 3. *P. pensylvanica*, Pin or Fire C. Lf., brt. and fr. nat. size; tip of brt. with terminal and uppermost lat. buds and lf. scars × 10. 4. *P. serotina*, Black C. Lf. and brt. nat. size; terminal and lat. buds × 5; lf. teeth enlarged. 5. *P. maritima*, Beach Plum. Lf., flowering (left) and vegetative brts. nat. size; lf. scar, lat. bud and part of brt. × 6.

The commonest cherry in our area is **P. serotina,** the Black or Wild Black C. Abundant as young growth along fences, and roadsides, one can readily recognize it from a distance in winter from its rather long, straight or wand-like shoots thrown off at a slight angle from the central axis.   In summer the thick shining lvs. with the callous tips of the teeth turning inward, are quite different from the dull, thinner lvs. of the

FIG. 68.   *Prunus glandulosa*, Flowering Almond.   Flowering brt. and lvs. nat. size.

Choke Cherry (**P. virginiana**), with their sharp saw-like teeth pointing outward.   The winter buds of that sp. are quite different from those of the Black C. being paler, with light margined scales, and the odor and taste of the twig is more bitter.   Further, the Choke-cherry is usually (though not always) shrubby and not at all as common as the Black C.

On higher elevations one finds the Pin C. with its long, narrow, finely serrate lvs. and small frs.; and almost anywhere, but never abundantly, especially near roads and houses, the Sweet Cherry, **P. avium,** is apt to occur, easily recognizable at any time of the yr. by its light colored, stout brts.   **P. cerasus,** the Sour C., is rare as an escape from cult.   In our area the Wild Plum, **P. americana,** readily distinguished by the flattened fr. pits, is occasionally seen.   The Peach, *P. persica*, is occasionally seen near roadsides, and readily distinguished by the reddened upper surface of the brts., green beneath.

The pink fls. of the Peach, opening in late Mar., are beautiful.   ⁰**P. maritima**, the Beach
Plum, is a shrubby sp. found only on or near sandy beaches along the coast but sometimes abundant there.   The fr. makes a delicious jelly or jam.

The best Japanese Flowering Cherries are mostly forms of *P. serrulata* (Fig. 69)
from Japan, China, and Korea, cult. for many centuries by the Japanese for their fls.

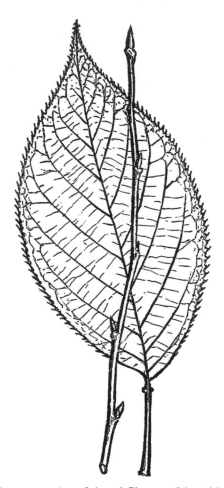

FIG. 69.   *Prunus serrulata*, Oriental Cherry.   Lf. and brt. nat. size.

One of the finest is var. Kwanzan, with double, rose-pink fls.   See P. & G. **6**: 33–36.
1950.   The display of this var. about May 1 at the Brooklyn Bot. Garden brings many
thousands of visitors each yr.   The forms at Washington, D. C., presented many yrs.
ago by the Japanese government, bloom about two weeks earlier and are mainly the
single vars.

LEGUMINÒSAE — PULSE FAMILY

Lvs. pinnate, except in *Cercis;* fls. papilionaceous, i. e. with parts as in a Sweet Pea fl., except in *Albizia, Gymnocladus* and *Gleditsia.* This is the Bean or Pea Family; fr. is therefore a legume like the bean or pea pod: *brts. and inner bark with taste of raw beans or peas;* true terminal bud usually lacking in our genera.

KEY TO GENERA

This key does not include *Cytisus, Genista,* or *Ulex*

ALBÍZIA

The Silk-tree, *Albízia julibríssin* Durazz., usually represented in our area by its hardier var. *rósea*, is sometimes cult. and rewards with its bright pink fls. in July and Aug., in heads, crowded at the upper ends of brts. Lvs. bipinnate, the lfts. very numerous and tiny—about ¼ in. long. A small tree, to about 35 ft. Also called Mimosa. Good specimens at Arnold Arboretum. Natzd. in se. States. Iran to c. China.

CÉRCIS — REDBUD

C. canadénsis L.   Redbud (Pl. XXI, 3).   Lvs. broadly heart-shaped, entire; buds small, blunt, glabrous, purplish, superposed and often collateral, with 2 or, in the fl. buds, with several scales exposed; lf. scars

FIG. 70. *Gleditsia triacanthos* and var. *inermis*, Honey-locust. Brts. with thorns and without thorns (var. *inermis*); bipinnate lf. × ½; center, tip of stem showing last lat. bud and, at left, stub of end of this year's growth × 5; left, above, lat. bud and lf. scar × 5.

somewhat raised, with decurrent ridges; bundle scars 3; stipule scars lacking; fls. pink, papilionaceous, borne close to old wood before lvs. appear; small tree.   Common wild around Washington, D.C.; e.g. in Rock Creek Pk.   S. Conn. and se. N.Y. to Fla. and w.

In the very similar oriental sp., *C. chinénsis* Bunge, Chinese R., sometimes cult., the lvs. have a very narrow, transparent margin, and the fls. are slightly larger (about ¾ in. long).

GLEDÍTSIA — HONEY-LOCUST

G. triacánthos L. Honey-locust (Fig. 70). Lvs. once or twice pinnate; pods a ft. or more long (sometimes less); fls. regular or nearly so; buds superposed, partly sunken in bark, but not so deeply as in coffee-tree; *thorns*, which are often *branched*, arise above the lf. axils (a thornless var. occurs, var. *inérmis*, see fig.): brts. *swollen* at or below lf. scars; bark comparatively smooth, but often with deep, more or less vertical fissures and many branched thorns. W. N.Y. to Fla. and w. Commonly cult. and established northeastward in N.E. to N.S.

GYMNÓCLADUS — COFFEE-TREE

G. dioìcus (L.) K. Koch. Coffee-tree (Pl. XXI, 2). Lvs. twice pinnate; dioecious; fls. regular; brts. stout, very irregular in arrangement and position; pith *salmon-colored;* buds two or three together, superposed, and deeply sunken in the bark; pods woody and thick, 5 in. or more long, rather persistent. Best known by its bark, which has thin, twisted ridges standing out at a wide angle from surface of trunk. Native from cent. N.Y. s. and w., but cult. n. to Can. Also known as Kentucky Coffee-tree.

CLADRÁSTIS — YELLOWWOOD

*C. lùtea (Michx. f.) K. Koch. Yellowwood (Pl. XXI, 1). Bark smooth and gray, like that of beech; buds naked, hairy, superposed, several so close together as to appear as a single bud, almost surrounded by lf. scar; stipule scars lacking; lvs. pinnate, petioles much swollen at base; fls. papilionaceous, white, fragrant, in June. Heartwood yellow. A rare tree in its wild state; much cult. Ky. and Tenn. to Ga. and w.

SOPHÒRA

S. japónica L. Japanese Pagoda-tree (Fig. 71). Brts. *dark green*, glabrous or slightly pubescent; buds small, reddish-brown, hairy, almost hidden under u- or v-shaped, raised lf. scars; stipule scars minute; lvs. pinnate; fls. very light yellow, sometimes with pink tinge, papilionaceous; in Aug. or Sept. China, Korea. Often cult.

LABÚRNUM

The Golden-chain, *L. anagyroìdes* Med. (Fig. 72), is a large shrub or small tree, to 20 ft., with yellow papilionaceous fls. in graceful, drooping racemes up to 1 ft. long, and 3-foliolate lvs. Buds with about 4 exposed, silvery-haired scales, lf. scars transversely elliptical, elevated, with 3 bundle scars close together. In *Cytisus* and *Genista* the lf. scars are minute, with 1 indistinct bundle scar.

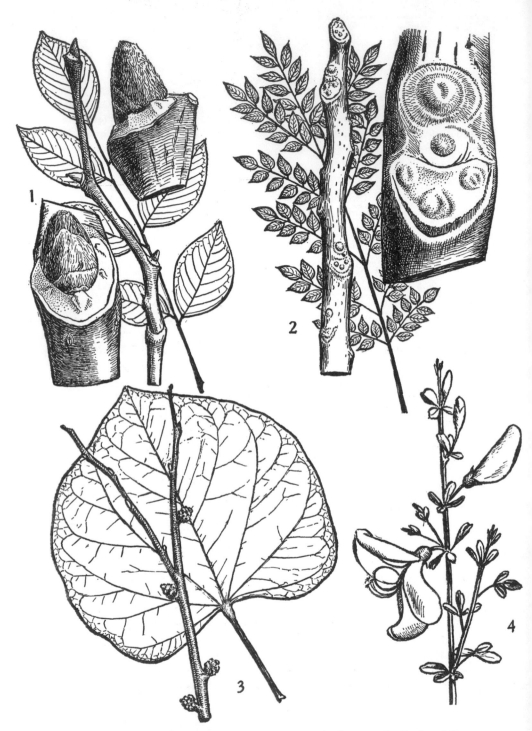

PLATE XXI. Pulse Family: Yellowwood, Coffee-tree, Redbud and Broom.

1. *Cladrastis lutea*, Yellowwood. Brt. nat. size; lf. × ⅛; buds, side and front views, and lf. scars × about 7.
2. *Gymnocladus dioicus*, Kentucky Coffee-tree. Brt. nat. size; lf. × ⅓; buds × about 5.
3. *Cercis canadensis*, Redbud. Lf. and brt. with fl. and lf. buds nat. size.
4. *Cytisus scoparius*, Scotch Broom. Flowering brt. nat. size.

Cýtisus — Broom

The Scotch Broom, [0]*C. scopàrius* (L.) Lk. (Pl. XXI, 4) with slender, green, ribbed or almost winged stems, 3-foliolate lvs., the upper lvs. often reduced to 1 lft., and fls. usually single or in 2's, is often cult. and sometimes natzd. in our area, especially southward. C. and s. Eu.

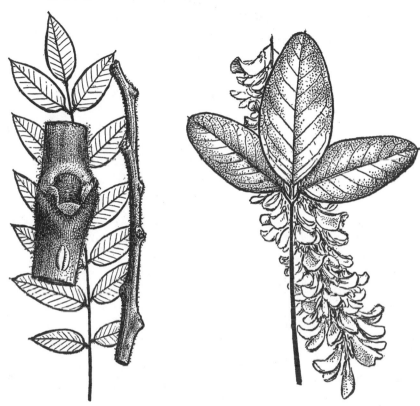

FIG. 71. *Sophora japonica*, Japanese Pagoda-tree. Lf. × ½; brt. nat. size; bud and part of brt. × 4.

FIG. 72. *Laburnum anagyroides*, Golden-chain. Lf. and fls. nat. size.

Genísta

A similar shrub is the Woadwaxen, Dyer's Greenweed, or Whin, [0]*Genísta tinctòria* L., which also has ribbed, but striped stems, usually lower—under 3 ft.—simple, ciliate lvs., and yellow fls. in many-flowered terminal racemes. Eurasia. Often cult., especially for rockeries. Rarely natzd.

Ùlex — Gorse

The Common Gorse or Furze, [0]*Ulex europaèus* L., in mature plants has no lf. blades, only triangular spiny petioles, and grooved, green brts. ending in sharp spines;

PLATE XXII.   Pulse Family concl.: Robinia and False-indigo.

1. *Robinia hispida*, Bristly Locust.   Brt., lf., fl. and young fr. nat. size.
2. *Amorpha fruticosa*, False-indigo.   Lfts. nat. size, showing stipels; lf. $\times$ ⅛; parts of brts.
   nat. size and $\times$ 4 to show variation in buds.
3. *Robinia pseudoacacia*, Black Locust.   Lf. and brts. nat. size; lf. scar with buried buds $\times$ 5;
   above, tip of brt., showing uppermost leaf scar and end of season's growth $\times$ 5.

fls. golden-yellow; buds small, superposed, i. e. 2 close together, appearing as 1. C. and w. Eu. Natzd. from se. Mass. southward, and often cult., especially for rockeries.

AMÓRPHA

⁰A. fruticòsa L. False-indigo (Pl. XXII, 2). Tall shrub to 12 ft., has pinnate lvs. marked with minute dots; spikes of violet, papilionaceous fls. in May and June, and 1–2-seeded, small, rough pods; buds with 4 or 5 scales exposed, superposed; stipule scars small. Native s. Pa. to Fla. and w. Cult. and escaped northeastward through N.E. and N.Y.

WISTÈRIA

Twining woody climbers, with pinnate lvs., and without tendrils or aerial rootlets; fls. blue, purple, or white, in racemes; lf. scars raised, and with horn-like protuberances at each side (at least on long shoots) which seem to be of assistance in climbing (Fig. 73).

⁰⁰*W. floribúnda* (Willd.) DC. Japanese W. One of most commonly cult. spp., with 11–19 lfts., callosities at base of standard, and *velvety* fr. pods. The fragrant, violet fls., in long, slender racemes (8–20 in.) open *gradually* from base to apex. Var. *macrobotrys* has fl. clusters 3 ft. or more long.

In the Chinese W., ⁰⁰*W. sinénsis* (Sims) Sweet, the lfts. are fewer (7–13), callosities present at base of standard, and pod also velvety, but in this sp. the fls. of one cluster (6–10 in.) all open nearly at the same time and are only slightly fragrant. The Nippon W., ⁰⁰*W. japónica* Sieb. and Zucc., has no callosities at base of standard, and white fls.; lfts. 9–13. The native sp., *⁰⁰W. frutéscens* (L.) Poir., Amer. W., has shorter, lilac-purple fl. clusters 1½–5 in. long, 9–15 lfts., and *smooth* fr. pods. Va. to Fla. and Ala. Without the fls. the spp. are difficult to name, but the genus

FIG. 73. *Wisteria floribunda,* Japanese Wisteria. Brt. with lat. bud showing lf. scar with "horns" at sides × 2.

can be easily recognized at any season by the peculiar backward-pointing hooks, sometimes only wart-like growths, at the lf. scars (Fig. 73) and by its twining character.

ROBÍNIA — LOCUST

R. pseùdoacàcia L. Black Locust (Pl. XXII, 3). Bark resembles somewhat that of Amer. elm; a pair of stipular thorns normally at nodes of brts., but often lacking (see fig.); buds superposed, hidden under fringed cracks of bark of lf. scar; lvs. pinnate; fls. white, papilionaceous, fragrant, in late May or June; pods 3–4 in. long, ½ in. wide, thin. Native from Pa. to Ga. and w., but thoroughly established in our area. A thornless form, var. *inérmis* exists; many other vars. cult. Introduced into France in the 17th century by Jean and Vespasian Robin and now natzd. over much of Eu. Valuable not only for ornament and very hard, durable wood, but, because of spreading, fibrous root system, to control erosion.

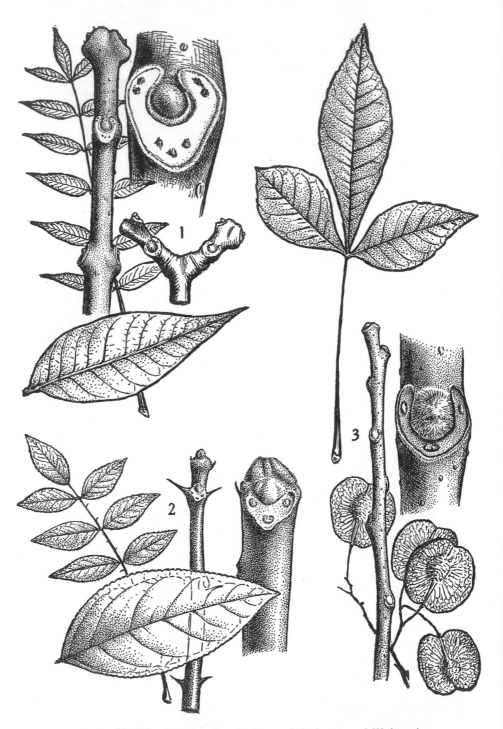

PLATE XXIII. Rue Family: Cork-tree, Prickly-ash and Wafer-ash.

1. *Phellodendron amurense*, Amur Cork-tree.   Brts., uppermost buds and lft. nat. size; lf. × ¼; lat. bud and lf. scar × 4.
2. *Xanthoxylum americanum*, Prickly-ash, Toothache-tree.   Brt. and lft. nat. size; lf. × ¼; tip of stem × 4.
3. *Ptelea trifoliata*, Wafer-ash.   Brt. and fr. nat. size;   part of brt. with lat. bud and lf. scar × 5; above, lf. nat. size.

The Clammy Locust, **R. viscòsa** Vent., with sticky, dark red-brown brts., and pink fls., is native from Pa. to Ga. and Ala.; cult.; often natzd. The Bristly L., or Rose-Acacia,* ⁰*R. hispida* L. (Pl. XXII, 1), with *bristly* brts. and rose-colored or pale purple fls., is a southern shrub (Va. and Tenn. s.), sometimes cult. and rarely escaped or natzd.

The Siberian Pea-tree, *Caragàna arboréscens* Lam., a rather large shrub or small tree with yellow, pea-like fls., green stems with vertical, corky ridges, 3 below each node, and stipules becoming thorny, is occasionally planted as an ornamental. Lvs. abruptly pinnate, each of 8–12 lfts., small and mucronate, the lvs. and fls. in close fascicles along brts. Very hardy. Siberia and Manchuria.

RUTÀCEAE — RUE FAMILY

Lvs. with pellucid dots, which are evident or easily seen with hand lens when lf. is held up to the light. These represent internal glands, containing a volatile oil, for this family is noted for its characteristic aromatic odors, e.g. in the Rue (*Rùta gravèolens*) a small subshrub, and notably in orange, lemon and other spp. of *Citrus*.

XANTHÓXYLUM — PRICKLY-ASH

(Also spelled with a "Z")

X. americànum Mill. Northern Prickly-ash, Toothache-tree (Pl. XXIII, 2). Rather rare shrub or small tree, to 25 ft.; brts. with a pair of stipular thorns at each node similar to those of the Black Locust. Lvs. alternate, pinnate with 5–11 lfts. Apt to be confused with the true ash, which, however, has opposite lvs. Has characteristic, lemon-like odor. Bark is medicinal. Que. to Ga. and w. Common in n. and w. N.Y.

PTÈLEA — HOP-TREE

P. trifoliàta L. Hop-tree, Wafer-Ash (Pl. XXIII, 3). Tall shrub or low tree; aromatic. Lvs. alternate, with 3 lfts., closely resembling those of poison-ivy, but the terminal lft. here is larger than the others, and *narrows gradually toward its usually sessile base* (in poison-ivy the terminal lft. is conspicuously stalked); also, the translucent dots characteristic of the lvs. of this family may readily be seen with a lens. Fr. a thin, circular disk composed of the fr. proper in the center surrounded by a circular wing, the whole about the size of a nickel; frs. in dense clusters persisting during the winter; buds very blunt, pubescent, almost surrounded or often more or less covered by the triangular, raised lf. scars; true terminal bud and stipule scars lacking. Fr. has been used as a substitute for hops. Bruised bark with strong odor. Que. to N.Y. to Fla. and w.

Phellodéndron — Cork-tree

The Amur Cork-tree, *P. amurénse* Rupr. (Pl. XXIII, 1) with *opposite*, pinnate lvs., keeled, silky-pubescent buds, and soft, corky, light gray bark, deeply grooved, is sometimes cult. and rarely natzd. True terminal bud rarely present; lf. scars horseshoe-shaped, almost surrounding buds; pith brownish; clusters of black, drupe-like frs. persistent during the winter; brts., except for their opposite lvs. and buds, similar in appearance to those of poison sumac. China.

SIMAROUBÂCEAE — QUASSIA FAMILY

Ailánthus

(Ai pronounced like "a" in paper)

A. altíssima (Mill.) Swingle (*A. glandulosa* Desf.) (Pl. XXIV, 1). Tree-of-heaven. Brts. thick (extremely so on young shoots), with a wide, brownish pith; lf. scars large, heart- or shield-shaped; buds hemispherical, more or less pubescent, relatively small, with 2 or sometimes 4 scales exposed, the terminal bud lacking; lvs. pinnate (1–2 ft. long), with a few blunt, glandular teeth at the base of each lft., and with a rank odor when crushed; *sap not milky*. Extensively natzd.; remarkably tolerant of city conditions, reproducing easily and growing vigorously in the most densely populated sections. China.

BUXÂCEAE — BOX FAMILY

Pachysándra

The Japanese Pachysandra, [0]*P. terminàlis*, Sieb. and Zucc. (Fig. 74), is a low, evergreen subshrub, often used as a ground cover; spreads by stolons, and is quite tolerant of shade.

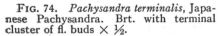

FIG. 74. *Pachysandra terminalis*, Japanese Pachysandra. Brt. with terminal cluster of fl. buds × ½.

FIG. 75. *Buxus sempervirens*, Common Box. Vegetative and fr. brts.

Búxus — Box

The Common Box, *B. sempervìrens* L. (Fig. 75), is usually a shrub with small elliptic or ovate, opposite, evergreen lvs. Cult. since ancient times, and much used for hedges. The True-Dwarf Box, var. *suffruticòsa* L., is used for edging of flower beds.

EMPETRÀCEAE — CROWBERRY FAMILY

The Black Crowberry, ⁰Èmpetrum nìgrum L., and the Broom Crowberry, ⁰Corèma conràdi Torr., are low, heath-like evergreen shrubs, with very small, linear lvs., the former with scattered and solitary fls. and small black, berry-like juicy drupes; the latter with fls. in terminal heads and small drupes, dry when ripe. The Black C. is found from Arctic regions to the Me. coast, alpine areas of N.E. and N.Y. and one station on L.I., N.Y. The Broom C. occurs from Nfd. s. along coast to se. Mass., mts. of N.Y., and N.J. Pine Barrens. E. nigrum is often planted in rock gardens where a dense mat of evergreen foliage is desired.

ANACARDIÀCEAE — CASHEW FAMILY

Rhùs — Sumac

(Pronounced "shoomac" or "soomac")

Sap milky (slightly so in R. copallìna); lvs. pinnate; lfts. serrate or entire; stipule scars lacking. The poisonous spp. of Rhus have whitish fr.; harmless, red. All except the fourth and sixth spp., although usually shrubs, sometimes attain the size and habit of small trees. Fall coloring of all spp. is magnificent.

According to SPN the poisonous spp. of Rhus are different enough to warrant their being placed in a separate genus, namely Toxicodendron. While I recognize the validity of this claim, I object to changing names which have long been in use. Prof. Fernald, in the 8th ed. of Gray's Manual of Botany (1950) retains the old names, although he places the poisonous spp. in the section "Toxicodendron."

Lvs. with 3 lfts.; low shrubs or creeping or climbing vines.
  Fr. red; fl. buds large, catkin-like, conspicuous in fall and winter; aerial rootlets
    absent; rather rare or local (Fig. 76) . . . . .4. R. aromatica 163
  Fr. white or nearly so; fl. buds not differentiated; brts. with tiny, aerial rootlets;
    very common; poison to touch (Pl. XXV, 2) . . . . .6. R. radicans 165
Lvs. pinnately compound, with many lfts.
  Brts. pubescent.
    Pubescence conspicuous; brts. and lvs. densely velvety-hairy; lf. axis (rachis)
      without winged extensions (Pl. XXV, 3) . . . . .1. R. typhina 161
    Pubescence not conspicuous; brts. soft-downy; lf. axis (rachis) with winged
      extensions (Pl. XXIV, 3) . . . . .3. R. copallina 163
  Brts. glabrous.
    Fr. red; brts. usually glaucous; lfts. serrate; shrub of dry soil (Pl. XXIV, 2)
                                       2. R. glabra 163
    Fr. white; brts. not glaucous; lfts. entire; tall shrub of wet, swampy soil; very
      poisonous to touch (Pl. XXV, 1) . . . . .5. R. vernix 163

1. R. typhìna L. Staghorn Sumac. (Pl. XXV, 3). Young brts. brown-hairy, somewhat resembling stags' antlers; buds almost surrounded by lf. scars; true terminal bud lacking; lfts. serrate; fr. red. Gaspé Pen. to Ga. and w.

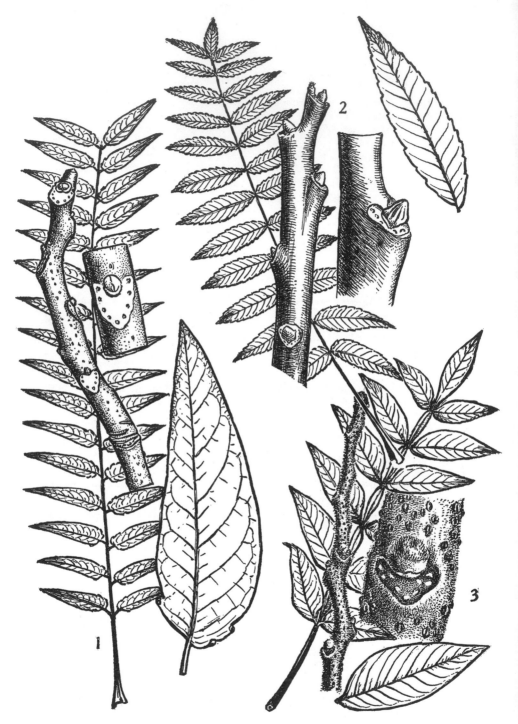

PLATE XXIV. Ailanthus and Sumacs.

1. *Ailanthus altissima*, Tree-of-heaven. Lft. and brt. nat. size; part of brt. with lf. scar and
  bud × 2; lf. × ⅓.
2. *Rhus glabra*, Smooth Sumac. Brt. and lft. nat. size; part of brt. with lat. bud × 2; lf. × ⅛.
3. *R. copallina*, Dwarf S. Brt. and lft. nat. size; part of brt. with lat. bud × 5; lf. × ½.

2. ⁰**R. glàbra** L. Smooth S. (Pl. XXIV, 2). *Brts. smooth*, usually glaucous; true terminal bud lacking; lfts. serrate; fr. red. Me. to Fla. and w. throughout the U.S.

3. **R. copallìna** L. Shining S. (Pl. XXIV, 3). Wing-rib or Dwarf S. Brts. downy; true terminal bud lacking; lvs. with *winged rachis*, lfts. mostly entire; sap watery; fr. red. Me. to Ga. and w. (See Gray, 8th ed., p. 977, for var. **latifòlia**).

FIG. 76. *Rhus aromatica*, Fragrant Sumac. Brt. showing fl. buds for next yr. and lvs. nat. size.

4. ⁰**R. aromática** Ait. Fragrant S. (Fig. 76). A prostrate shrub, resembling Poison-Ivy with its 3 lfts., but they are pubescent and fragrant when bruised and the fr. is *red*. Juice watery but gummy. Lfts. dentate above lowest third. Peculiar in its large scaly fl. buds conspicuous in fall and winter. Not a common plant, but locally abundant. "Rather common in the Hudson valley" (House (28) 1924). For vars. see Gray, 8th ed. p. 978. Que. to Fla. and w. (*R. canadénsis* Marsh.)

5. **R. vérnix** L. Poison S., Poison-elder, or Poison-dogwood (Pl. XXV, 1 and fig. 2). Brts. smooth, speckled with dark dots (lenticels);

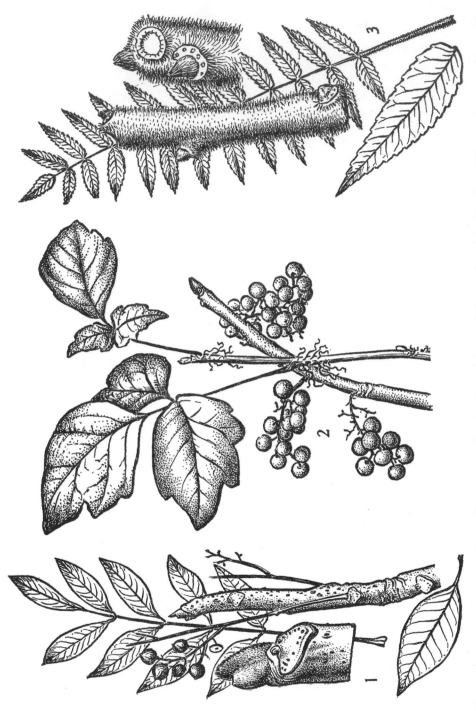

PLATE XXV. Sumacs concl. 1. *Rhus vernix*, Poison S. Brt. with part of cluster of grayish-white fr. and lft. nat. size; apex of brt. with end bud and lf. scar × 3; lf. × ½. 2. *R. radicans*, Poison-ivy. Brts. and lvs. showing fr. (whitish) and rootlets, nat. size. 3. *R. typhina*, Staghorn S. Brt. and lft. nat. size; apex of brt. showing buds, lf. scars and terminus of year's growth × 2; lf. × ⅓.

true terminal bud present; lfts. entire; fr. whitish; grows only in wet places. *Very poisonous to the touch.* Que. to Fla. and w.

6. ⁰⁰**R. radìcans** L. Poison-ivy, "Poison-oak" (Pl. XXV, 2 and fig. 2). A woody climber or vine with aerial rootlets, but often creeping on the ground and sending up short, erect shoots; buds stalked; fr. glabrous, whitish; lfts. 3, entire or toothed, with stalks. *Poisonous to the touch.* Cf. *Ptelea*, p. 159. *Staphylea*, p. 170, also has 3 lfts., but there the buds are opposite. Que. to Fla. and w.

The lfts. in Poison-ivy are very variable, sometimes almost entire, sometimes deeply serrate and/or lobed as in Pl. XXV, 2, and the lobed or deeply cleft form has probably led to the common name "Poison-Oak." However, the true Poison-oak, ⁰R. toxicodéndron L., has no aerial rootlets, does not climb, has mostly simple, erect stems, to 2 ft., with lfts. *tomentose* beneath, and has a southerly range, from N.J. to Fla. and w. (See Gray's Manual; 8th ed. p. 979).

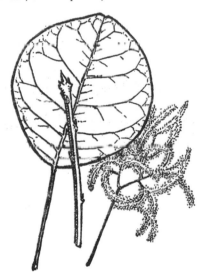

Fig. 77. *Cotinus coggygria*, Smoke-tree. Lf., brt. and plumose fr. stalks nat. size.

CÓTINUS — SMOKE-TREE

The Smoke-tree, *C. coggýgria* Scop., (Fig. 77), is closely related to the Sumacs, but has simple lvs. with the fr. stalks becoming plumose after flowering, giving rise to the common name. Much cult. Eurasia.

AQUIFOLIÀCEAE — HOLLY FAMILY

ÌLEX — HOLLY

Fls. small, in the lf. axils, in June, usually dioecious; fr. a berry-like drupe. Therefore, in order to set fr. a pistillate plant must be reasonably

PLATE XXVI. Hollies.

1. *Ilex opaca*, Amer. Holly.   Brt. and lvs. nat. size.
2. *I. glabra*, Inkberry.   Brt. with lvs. and fr. nat size.
3. *I. verticillata*, Common Winterberry.   Below: lf., brt. and fr. nat size; above: tip of brt. × 4.
4. *I. crenata*, Japanese H.   Brt. and lvs. nat. size.
5. *Nemopanthus mucronata*, Mountain-holly.   Brt., lvs. and fr. nat size.

near one or more staminate plants. The pollen from these plants may then reach the fls. of the pistillate plant, effecting fertilization and the development of the berries. The staminate plants, with occasional exceptions, do not bear fr.

As a matter of fact the situation is somewhat complicated, because, at least in some spp., the fls. of the staminate and pistillate plants contain rudimentary organs of the opposite sex and these organs sometimes develop to maturity.

1. **I. opàca** Ait. Amer. Holly (Pl. XXVI, 1). Lvs. persistent, spiny, not so glossy nor with such wavy margins as in English H.; *berries red*. Common southward not far from the coast but extends inland to Mo. and Okla.; the holly commonly used here for Christmas decoration. E. Mass. to Fla. and w.

2. **I. verticillàta** (L.) Gray. Common Winterberry (Pl. XXVI, 3). Lvs. deciduous, serrate, *dull* above, highly variable in shape, not spiny; *berries red* and much the same as in Amer. H.; brts. very slender, dark purple or grayish purple; buds tiny, blunt, single or often superposed, a small one very close to the base of a larger one; lf. scars crescent-shaped, raised, with one vascular bundle scar; *true terminal bud* and *minute stipule scars or remains of tiny, pointed stipules present*. Also popular for decoration, and becoming scarce; moist soil or swamps. Nfd. to Ga. and w. Several vars. common, especially var. *padifolia*. See Fernald (14), pp. 981, 982.

The Smooth Winterberry 3. I. **laevigàta** (Pursh) Gray, has shining, more finely toothed lvs., glabrous beneath, or hairy only on the veins, and grows in somewhat drier soil. S. Me. to Ga.

The Mountain Winterberry, I. **montana**, T. and G., a large shrub or small tree, is a more upland sp. Lvs. *sharply serrate*, 2–6 in. long, and drupes large—about 2/5 in. (1 cm.) in diam. Rich, wooded slopes, e. to w. N.Y. and s. upland to Ala. and Tenn. Rather rare.

4. ⁰**I. glàbra** (L.) Gray. Inkberry (Pl. XXVI, 2). Smooth, shining, evergreen or half-evergreen lvs., almost entire, and *black fr.* N.S. to Fla. and w. Often cult.

The English H., 5. *I. aquifòlium* L., is hardy only in the extreme southern part of our area. At Brooklyn, N.Y., in severe winters it is killed back. Distinguished by its *glossy*, evergreen lvs., with *wavy*, spiny margins, and red berries. Cult. since ancient times in many vars. Eu., N.Afr., and Asia.

The Japanese H., 6. ⁰*I. crenàta* Thunb. (Pl. XXVI, 4), with small, shining, crenately toothed (at least in outer half) evergreen lvs., borne very thickly, and *black fr.*, is much cult. and thrives under city conditions. Not very hardy in our area; at Brooklyn, N.Y. sometimes suffers in severe winters.

NEMOPÁNTHUS — MOUNTAIN-HOLLY

⁰**N. mucronàta** (L.) Trel. Mountain-holly (Pl. XXVI, 5). Gray-barked shrub to 9 ft., with simple, deciduous, entire or slightly toothed

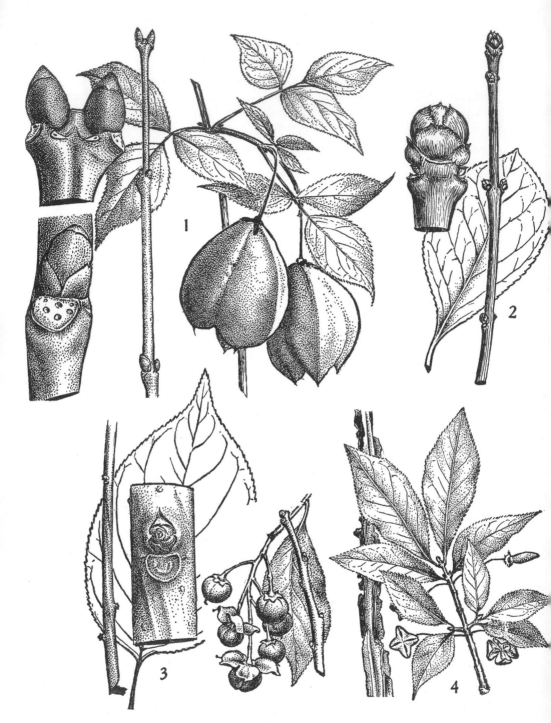

PLATE XXVII.   Euonymus, Climbing Bittersweet and Bladdernut.

1. *Staphylea trifolia*, Amer. Bladdernut.   Brt. with leaves and fr. nat. size; winter brt. nat. size; lat. ar
   uppermost buds × 5.
2. *Euonymus europaeus*, European Euonymus, Eu. Spindle-tree.   Lf. and brt. nat. size; end of brt. wi
   terminal bud × 5.
3. *Celastrus scandens*, Climbing Bittersweet.   Lf. and brt. nat. size; part of brt. with lat. bud and lf. scar
   about 5; rt., fr. and brt., nat. size.
4. *Euonymus alatus*, Winged E., Winged Spindle-tree.   Brts. with lvs. and fr., and with corky wings na
   size.

lvs.; conspicuous in the northern woods in summer and fall from its bright red, long-stalked fr., about the size of a pea, borne in fair abundance throughout the whole plant. Damp woods and swamps. Nfd. to upland Va. and w.

CELASTRÀCEAE — STAFF-TREE FAMILY

Euónymus

E. europàeus L. European E., Spindle-tree (Pl. XXVII, 2). Opposite, simple, crenately toothed lvs.; small, conical or rounded buds; slender, *green brts.* with corky lines on the angles; fr. 4-lobed, smooth, pink, with red inner parts disclosed when ripe. Small tree, often cult. and sometimes natzd. Eurasia.

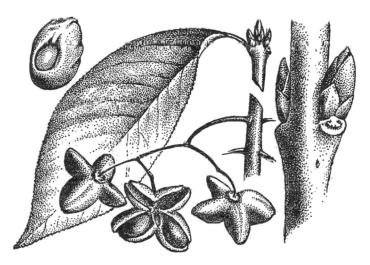

Fig. 77a. *Euonymus atropurpureus*, Jacq. Burning-bush. Upper left, aril, open showing part of seed; parts of brts. showing lf., terminal bud group, and frs.; right, part of brt. with lat. buds and lf. scar enlarged.

Amer. spp. of E. occasionally seen are: **E. atropurpùreus** Jacq., Burning-bush (Eastern Wahoo SPN), with 4-parted *purple* fls., and lvs. pubescent beneath; Ont. to Ala. and w. (Fig. 77a); and ⁰**E. americànus** L. Strawberry-bush, with 5-parted fls., pink, *warty* fr., scarlet inside, both spp. without lines or corky ridges on stems. Se. N.Y. to Fla. and w.

The Winged E., ⁰*E. alàtus* (Thunb.) Sieb. (Pl. XXVII, 4), with corky wings on brts., a shrub to 9 ft., is cult., especially in var. *compáctus*, in which lvs. turn a striking crimson in Oct. Asia.

The Winterberry Euonymus, *E. bungeanus* Maxim., with abundant, deeply 4-lobed, pink or yellowish, *smooth* fr., long persistent into winter, is a shrub or small tree to 18 ft., a popular and desirable ornamental. The lvs. are elliptic-ovate to elliptic-lanceolate,

2–4 in. long, very long acuminate and serrate with a long petiole.   At fl. time the anthers are purple.   Seeds white or pinkish with orange aril, usually open at end.   N. China, Manchuria.

The Winter-creeper E., ⁰⁰*E. fortùnei* var. *radìcans* (Miq.) Rehd., with evergreen lvs. and climbing by aerial rootlets, is often cult.   Excellent both as ground cover and for covering walls.   The sp. is Chinese.

CELÁSTRUS — BITTERSWEET

⁰⁰**C. scándens** L.   Climbing Bittersweet (Pl. XXVII, 3).   A woody *stem twiner*, with orange-colored, berry-like fr., showing scarlet interior when ripe; lvs. simple, finely serrate, alternate; buds small, projecting like small knobs at right angles to stem.   Que. to Ga. and w.

The Oriental B., ⁰⁰*C. orbiculàtus* Thunb., is similar, but has more rounded lvs., crenately toothed, and fls. and frs. in small *axillary* clusters (in **C. scandens** in terminal clusters).   A vigorous grower; sometimes planted, *e.g.* along Hutchinson R. Pkway. Westchester Co., N.Y.   Occasionally escaped.   Japan and China.

STAPHYLEÁCEAE — BLADDERNUT FAMILY

STAPHYLÈA — BLADDERNUT

⁰**S. trifòlia** L.   Amer. B. (Pl. XXVII, 1).   Bark striped; buds opposite, with 3 or 4 scales showing, smooth, red-brown, pointed; lf. scars triangular, with 3–6 bundle scars; stipule scars present; terminal bud absent; lvs. with 3 finely serrate lfts.; fr. a 3-divided bladdery pod.   Que. to Ga. and w.   Should be more planted as an ornamental shrub.

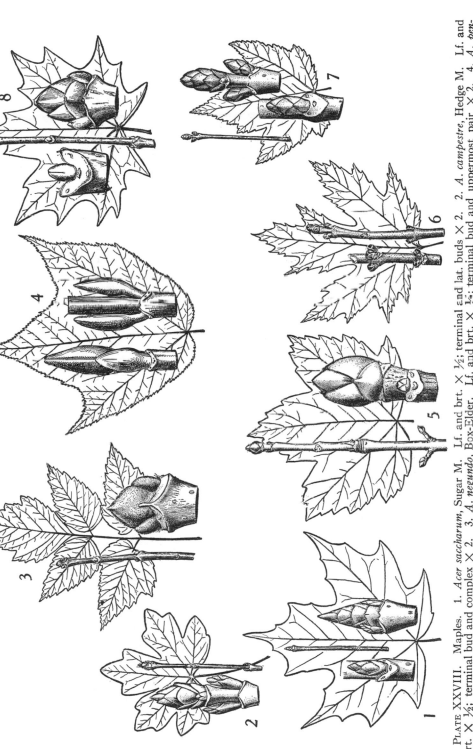

PLATE XXVIII. Maples. 1. *Acer saccharum*, Sugar M. Lf. and brt. × ½; terminal and lat. buds × 2. 2. *A. campestre*, Hedge M. Lf. and brt. × ½; terminal bud and complex × 2. 3. *A. negundo*, Box-Elder. Lf. and brt. × ½; terminal bud and uppermost pair × 2. 4. *A. pensylvanicum*, Striped M. Lf. × ½; terminal and one of uppermost lat. buds and pair of lat. buds × 2. 5. *A. pseudoplatanus*, Sycamore M. Lf. and brt. × ½; terminal bud and tip of brt. × 2. 6. *A. saccharinum*, Silver M. Lf. and brts. × ½. 7. *A. rubrum*, Red M. Lf. and brt. × ½; tip and lower portion of brt. showing buds and lf. scars × 2. 8. *A. platanoides*, Norway M. Lf. and brt. × ½; terminal and lat. buds × 2.

ACERACEAE — MAPLE FAMILY

Acer — Maple

Lvs. in our spp. simple (except pinnately compound in *A. negundo*), palmately lobed, opposite; true terminal bud present, except usually in *A. palmatum* and *japonicum;* lf. scars triangular, u-shaped or linear; bundle scars in 3 groups; stipule scars lacking; fr. *a double samara.*

Lvs. compound, with 3–5 lfts.; buds pubescent; brts. glaucous (Pl. XXVIII, 3)
        7. *A. negundo* 174
Lvs. simple, lobed.
    Lvs. with milky juice (shown by breaking lf. stalk, or cutting across a bud scale.)
        Lvs. 2–4 in. across, pubescent beneath, lobes with rounded teeth (Pl. XXVIII, 2)..6. *A. campestre* 174
        Lvs. averaging larger, 4–7 in. across, glabrous beneath and shining, lobes with pointed teeth (Pl. XXVIII, 8).......................2. *A. platanoides* 173
    Lvs. without milky juice.
        True terminal bud absent; brts. ending with the last 2 laterals.
        Japanese Maples.
            Lvs. deeply 5–9 lobed; petioles glabrous (Fig. 78)........10. *A. palmatum* 175
            Lvs. not so deeply 7–11 lobed; petioles pubescent (Fig. 79).11. *A. japonicum* 175
        True terminal bud present; brts. ending with a terminal bud and usually a lateral bud on each side.
        Buds with only 2–4 scales showing; lvs. mostly 3-lobed.
            Brts. green, glabrous; bark of trunk with white streaks; buds stalked, bud scales 2, red, glabrous, valvate (Pl. XXVIII, 4).8. *A. pensylvanicum* 175
            Brts..red-brown, downy, at least near tip, often streaked with green; buds slightly stalked, 2–4 scales showing, the inner pair tomentose (Fig. 79a)
                9. *A. spicatum* 175
        Buds with many scales showing (more than 4); lvs. 3–5 lobed.
            Buds sharp-pointed (Pl. XXVIII, 1)..................1. *A. saccharum* 172
            Buds blunt.
                Buds usually green, sometimes with pink or reddish tinge; frs. in long pendulous clusters (Pl. XXVIII, 5)...........5. *A. pseudoplatanus* 174
                Buds reddish; lf. sinuses mostly sharp-pointed.
                    Lvs. deeply cleft, silvery-white beneath (Pl. XXVIII, 6)
                        3. *A. saccharinum* 173
                Lvs. not so deeply cleft.
                    Lvs. pale or glaucous beneath; native (Pl. XXVIII, 7)
                        4. *A. rubrum* 173
                    Lvs. light green beneath; cult (Fig. 79b), p. 174....12. *A. ginnala* 175

1. **A. sáccharum** Marsh.   Sugar Maple (Pl. XXVIII, 1).   Bark gray; not scaly; in old trees in long, thick, irregular plates; *buds sharp-pointed and scaly*, about ¼ in. long, brown or often purplish, somewhat pubescent, especially toward tip; lvs. with *rounded sinuses* and sparingly toothed, *pale and glabrous beneath.*   Gaspé Pen. to Del., upland to Ga. and w.

The Black M., **A. nìgrum** Michx. f., strongly resembles Sugar M., with about same range, but lvs. are pubescent beneath and deeper green; brts. are hairy when young, later glabrous, orange or yellow all yr. through; bark of old trunks black; lvs. mostly 3-lobed. Que. to Ga. and w.

2. **A. platanoìdes** L. Norway M. (Pl. XXVIII, 8). Bark close (not scaly); *buds large, reddish* (sometimes intermixed with green); *sap milky;* lvs. like those of Sugar M. but green and shining beneath. A red-lvd. var. commonly cult. is var. *schwédleri.* The red color disappears in late June. Eu., Caucasus.

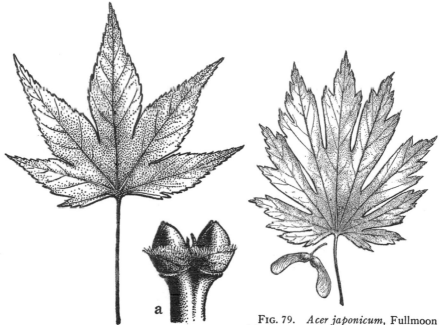

FIG. 78. *Acer palmatum*, Japanese Maple. Lf. nat. size; a, winter buds at end of brt. × 6.

FIG. 79. *Acer japonicum*, Fullmoon M., Japanese Maple. Lf. and fr. × ½ (one of many forms; lf. lobes often not so much toothed).

3. **A. sacchariìnum** L. Silver M. (Pl. XXVIII, 6). Bark scaly; buds small, red; lvs. deeply cleft, silvery white beneath, with sinuses sharp, or more or less so. Brts. often pendulous, with rank odor when crushed, and tend to point upward at their tips. Moist soil and along streams; rare near coast. N.B. to Fla. and w.

Wier's Cutleaf M., *A. sacchariìnum* var. *wìeri*, is a form of *A. sacchariìnum laciniàtum* (Carr.) Pax. with very deeply cleft lvs. and narrow lobes.

4. **A. rùbrum** L. Red M. (Pl. XXVIII, 7). Bark scaly, much like that of the last, but on young trees and brs. pale as in the beech; buds

similar to those of last sp.; lvs. not so deeply cleft, but always with *sharp sinuses*, glaucous beneath. The common wild sp. of our area; usually in low ground or swamps, sometimes in drier soil. Nfd. to Fla. and w. Also commonly known as Swamp or Soft M.

5. **A. pseudoplátanus** L. Sycamore M. (Pl. XXVIII, 5). Bark in roundish, irregular, exfoliating scales; buds *green*, sometimes tinged with red; lvs. with *sharp sinuses;* lf. veins prominent on lower surface and pubescent; *fls. and frs. in long pendent clusters.* Eurasia.

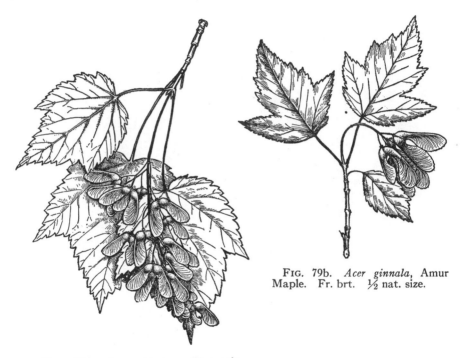

Fig. 79b. *Acer ginnala*, Amur Maple. Fr. brt. ½ nat. size.

Fig. 79a. *Acer spicatum*, Mountain Maple. Fr. brt.; topmost lf. turned over showing pubescent undersurface. ½ nat. size.

6. A. campéstre L. Hedge M., European Field M. (Pl. XXVIII, 2). Bark close; *sap milky;* lobes and teeth of lvs. rounded. Eurasia.

7. **A. negundo** L. Box-elder, Ashleaf M. (Pl. XXVIII, 3). Buds short-stalked, reddish and usually woolly; lf. scars v-shaped; *brts.* (usually glaucous) *green or reddish; lvs. compound*, with usually 3–5 lfts.; fr. in long (6 in.) pendulous clusters. Very hardy and drought-resisting. W.N.E. to Fla. and w. Much cult. and natzd. in n. N.E. and e. Can.

8. **A. pensylvánicum** L.   Striped M., Moosewood (Pl. XXVIII, 4).
Small tree to 35 ft. with smooth, green, white-striped bark; lvs. 3-lobed,
large, finely serrate; buds red, with 2 valvate scales showing, terminal bud
stalked and large.   Que. to Ga. and w.

9. **A. spicàtum** Lam.   Mountain M. (Fig. 79a).   Mostly shrubby;
bark thin, red-brown; lvs. pubescent beneath, mostly 3-lobed, coarsely
serrate; buds small, with short stalks, 2–4 scales showing, terminal bud
small, about ¼ in. long, including stalk.   Cool woods in n. Nfd. to n.
N.J., Pa. and w.

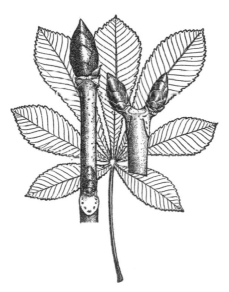

FIG. 80. *A e s c u l u s hippocastanum*,
Horse-Chestnut. Lf., brt. with terminal
bud and brt. with 2 lat. buds (having
ended in a fl.) × ½.

FIG. 81.   *Aesculus
carnea*, Red Horse-
Chestnut. Brt. with
terminal bud nat. size.

Two small spp. of Japanese Maples are commonly cult. (some of their vars. shrubby): (10) the Japanese M., *A. palmàtum* Thunb. (Fig. 78), with lvs. very deeply palmately 5–9-lobed (a common var. has red lvs.); and (11) the Fullmoon M., *A. japónicum* Thunb. (Fig. 79), with lvs. not deeply 7–11-lobed.   In both of these the terminal
bud is usually lacking.   (12) The Amur maple, *A. ginnala* Maxim. (Fig. 79b) from C.
and N. China, Manchuria, and Japan is rarely planted.   A small tree, to 20 ft., with
3-lobed, doubly serrate lvs., glabrous and green beneath, the *middle lobe much larger*
and often asymmetrical.   Various lf. forms occur.   Lvs. right red in fall.   Fr., including wing, about 1 in. long, rosy during summer.   Brts. glabrous but warty.   Very
hardy.   Good specimen in Berkeley College Campus, Yale Univ.

HIPPOCASTANÀCEAE — HORSE-CHESTNUT FAMILY

AÈSCULUS — HORSE-CHESTNUT

A. hippocástanum L.   Horse-chestnut.   (Fig. 80).   Buds large, resinous, opposite; *lvs. opposite, palmately compound*, with usually 7 lfts. Fls. in large, erect, cone-shaped clusters, through whole tree; fr. prickly. Much cult. and sometimes natzd.   Balkan Peninsula.

Large, pink- or red-flowered Horse-chestnuts, often cult. are usually *A. cárnea* Hayne (Fig. 81), with *slightly resinous* buds, 5 lfts. and somewhat prickly fr., a cross between **A. hippocastanum** and **A. pàvia* L., the latter a shrubby, red-flowered sp. of the southeastern U.S.

The Ohio Buckeye, **A. glàbra* Willd., a native of the Middle West, sometimes cult. in our area, has non-resinous buds, usually 5 lfts. and prickly fr.

The Yellow B., **A. octándra** Marsh, a tall tree to 90 ft. with usually 5 (sometimes 7) lfts. and usually 2-seeded fr. without prickles, is sometimes cult.   W.Pa. to Ga. and w.

SAPINDÀCEAE — SOAPBERRY FAMILY

KOELREUTÈRIA — GOLD-RAIN-TREE

The Gold-rain-tree, also called China-tree, *Koelreutèria paniculàta* Laxm., to 30 ft., is conspicuous with its panicles of bright yellow fls. in summer; lvs. pinnate or bipinnate, 1 ft. or more long, with lfts. which are serrate and lobed toward base; fr. bladdery, 3-celled, somewhat similar to that of its relative, the Bladdernut (p. 170). Occasionally cult.   China, Korea, Japan.

RHAMNÀCEAE — BUCKTHORN FAMILY

RHÁMNUS — BUCKTHORN

Shrubs or small trees with alternate or opposite simple lvs. and black, berry-like fr.

The Glossy B., **Rhamnus frángula** L., is a shrub or small tree to 18 ft. with alternate, oblong to obovate lvs. 1–3 in. long; small naked buds and dark purple fr. in small sessile umbels.   Natzd. from N.S. to N.J. and w.   Eurasia, n. Afr.   Other spp. in our area are **R. cathártica** L., the European B. (Fig. 82) with opposite lvs., natzd., from Eu. and Asia, and ⁰**R. alnifòlia** L'Her., Alder B.   Cascara Sagrada, the well-known laxative, is prepared from the bark of **R. purshiàna* DC., which grows in northwest U.S. and adjacent Canada.

CEANÒTHUS

The New-Jersey-tea, ⁰**C. americànus** L., grows up to 3 ft. and has small hairy buds; lvs. alternate, ovate, smooth or slightly pubescent below, acute at tip, shallowly toothed, with 3 prominent nerves and small deciduous stipules. Que. to Fla. and w. Lvs. used for tea during Amer. Revolution.

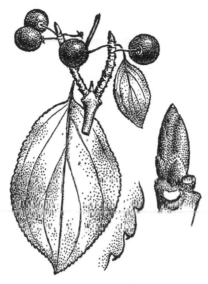

Fig. 82. *Rhamnus cathartica*, European or Common Buckthorn. Brts., lvs., and fr. nat. size; terminal bud of short growth × 5; lf. teeth × about 7.

VITÂCEAE — GRAPE FAMILY

Vìtis — Grape

Lvs. simple, usually lobed; pith brown, usually with a diaphragm at each node; stems striate; tendrils branched or simple without disks at their tips.

Two spp. of wild grape are common in our area. Fox G., ⁰⁰V. labrúsca L. (Fig. 83), has very woolly brts.; lvs. persistently woolly below; a tendril or a fl. cluster at every node; and large, dark purple or amber-colored frs.; has given rise to the Concord and many other vars. of cult. grapes. S. Me. to Pa. and upland to Ky. and Tenn.; rare on Coastal Plain to Ga.

The Summer G., ⁰⁰V. aestivàlis Michx., has more loosely pubescent brts.; mature lvs. with scattered rusty woolliness below; tendrils intermittent; and much smaller black frs. with a bloom. Mass. to Ga. and w.

Ampelópsis

The Porcelain-berry, ⁰⁰A. brevipedunculàta (Maxim.) Trautv., with 3-lobed, alternate, serrate lvs. and bright blue fr. in clusters, climbs by tendrils and is sometimes cult. for its ornamental, turquoise-colored fr. (Amur ampelopsis SPN). Ne. Asia.

Parthenocíssus

⁰⁰P. quinquefòlia (L.) Planch. Virginia Creeper (Pl. XXX, 2). Sometimes mistaken for Poison-ivy, but can be distinguished by its

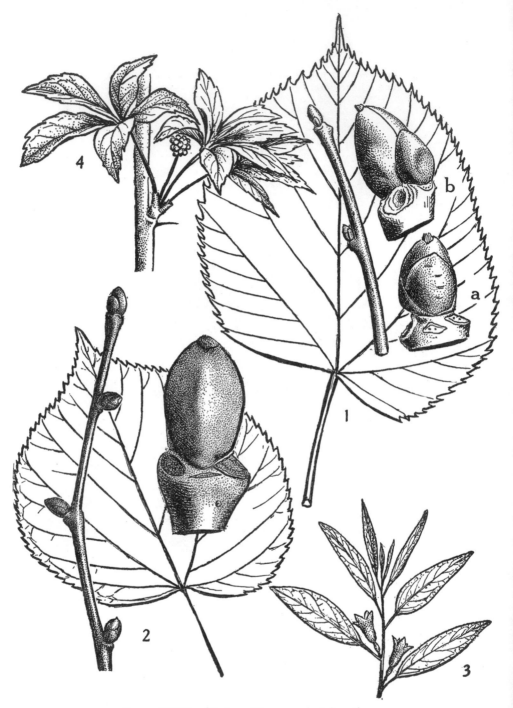

PLATE XXIX. Lindens, Elaeagnus and Acanthopanax.

1. *Tilia americana*, Amer. Basswood, Linden. Lf. and brt. nat. size; uppermost buds × 4;
   a, at base, left to rt., end of season's growth (profile view), stipule scar, lf. scar; b, at base:
   end of season's growth (surface view), lf. scar and smaller, accessory bud.
2. *Tilia europaea*, Eu. L.   Lf. and brt. nat. size; uppermost lat. bud and end of brt. showing, at
   base, left to rt., end of brt., stipule scar, and lf. scar × 4.
3. *Elaeagnus umbellata*, Autumn Elaeagnus.   Brt. with lvs. and fls. nat. size.
4. *Acanthopanax sieooldianus*, Acanthopanax.   Brt. showing fascicle of lvs., fl. buds and thorn
   nat. size.

palmately compound lvs. of 5 lfts.; bluish black berries; circular, raised lf. scars subtending blunt buds (often 2 at a node, 1 large and 1 small); *and by its usual lack of aerial rootlets,* the branched tendrils with expanded adhesive disks at their tips serving as holdfasts (older stems sometimes develop aerial roots in abundance); pith white or greenish.   Que. to Fla. and w.   Also called Woodbine.

FIG. 83.   *Vitis labrusca,* Fox Grape.   Lf. and brt. × ½; part of brt. with lf. scar and lat. bud × 2½.

In the Boston-ivy, ^{oo} *P. tricuspidàta* (Sieb. and Zucc.) Planch. (Pl. XXX, 1), the lvs. of basal shoots are mostly composed of 3 lfts.; others are mainly 3-lobed.   Much cult. for covering walls; clinging by branching tendrils with flattened disks, as in Virginia Creeper; lvs. bright scarlet or orange in fall; fr. blue-black; sometimes escaped from cult.; hardy except in extreme northern N.E.; several cult. vars.   The true Ivy, *Hédera,* which is evergreen, belongs to a different family (p. 188 and fig. 91).   It is possible to confuse the form with 3 lfts. with Poison-ivy, but there the stems have *aerial rootlets,* no tendrils, and fr. is whitish.

TILIÀCEAE — LINDEN FAMILY

TÍLIA — LINDEN or BASSWOOD

Stalk of fl. or fr. cluster appears to grow from about the middle of a large strap-shaped bract; fr. hard, spherical, about the size of a pea; lvs. simple, alternate, more or less heart-shaped; buds lopsided, 2 or 3 scales exposed; true terminal bud lacking; stipule scars unequal.   The heart-shaped lf. and lopsided bud are very distinct characters.

T. americàna L.   Amer. Basswood, Amer. Linden (SPN) (Pl. XXIX, 1).   Lvs. heart-shaped, large (3–8 in. long), the under surface with tufts of hairs in axils of lateral veins but *wanting in those at base of lf.;* buds carmine or greenish, with the characteristic large scale on one side, giving the usual lopsided appearance.   N.B. to N.C. and w.

A similar Amer. sp. is **T. neglécta** Spach., with lvs. grayish and loosely stellate-pubescent beneath.   Que. to N.C. and w.   In **T. heterophýlla** Vent., the White Bass-wood or Bee-tree Linden, the lvs. are white beneath with crowded stellate pubescence. N.Y. to Fla. and w.

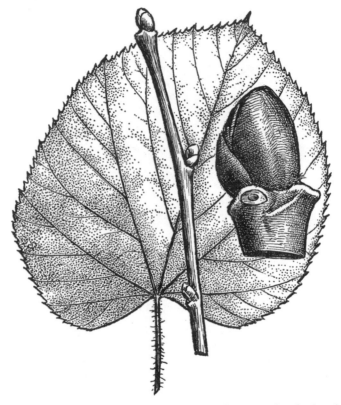

FIG. 84.   *Tilia tomentosa*, Silver Linden.   Lf. and brt. nat. size; bud and tip of brt. × 5.

A common European sp., much cult., is the European L., *T. europaèa* L. (Pl. XXIX, 2), with smaller lvs., which have *tufts of hairs in axils of all the veins.*   Baxter Blvd., Portland, Me., is lined on both sides with *T. europaea.*   The Silver L., of se. Eu. and w. Asia, *T. tomentòsa* Moench. (Fig. 84), occasionally cult., is easily recognized by the white-tomentose under surface of the sometimes doubly serrate lvs. and by the pubescent petioles.

T. cordàta Mill.   Small-lvd. Eu. L. or Little-leaf L., recognizable from its small heart-shaped lvs., is also much cult.   Eu.   Occasionally one sees the Large-lvd. L., *T. platyphýllos* Scop., a Eu. sp. with large lvs. (2½–5 in. long), green-velvety beneath, and sometimes also above.

MALVÀCEAE — MALLOW FAMILY

Hibíscus

The Shrubby Althea or Rose-of-Sharon, ⁰*H. syríacus* L. (Fig. 85), valued for its late blooming fls. (Aug. and Sept.) is much cult. and sometimes escaped. Lvs. 2–4 in. long, ovate or rhombic ovate, more or less 3-lobed with rounded or acutish teeth. Buds hidden above lf. scars, twigs rounded, fluted near dilated tip; fls. white to red or purple or violet, double forms frequent. Fr. a capsule, splitting open when ripe. China, India.

FIG. 85. *Hibiscus syriacus*, Shrubby Althea, Rose-of-Sharon. Fl., fr., lvs., and buds × ½.

ACTINIDIÀCEAE — ACTINIDIA FAMILY

Actinídia

The Bower Actinidia, ⁰⁰*A. argùta* (Sieb. and Zucc.) Miq. (Fig. 86), is a hardy, high-climbing woody vine; brs. glabrous with brown, chambered pith; buds buried in a peculiar swelling above lf. scar; no true terminal bud; lvs. rather thick, broad-ovate to elliptic, abruptly acuminate, serrate, the serrations with bristle-like teeth; pith chambered with very thin, brownish, often forked plates (in section); fr. globular or ellipsoid, greenish-yellow, about 1 in. long. Very hardy. Fr. edible, resembling a grape but with small seeds. Deserves to be better known. Occasionally cult. Japan, Korea, Man-churia.

THEÀCEAE — TEA FAMILY

Franklínia

The Franklinia, *F. alatamàha* Bartr., is a shrub or small tree to 30 ft.; with large obovate-oblong lvs. (5–6 in. long) gradually narrowed into the petiole, shining above, pubescent beneath; buds round-ovoid, *naked*; fls. large, white, about 3 in. in diam., in Sept. and Oct. Discovered in 1765 near Ft. Barrington, McIntosh Co., Ga., by John and William Bartram. Not found wild since 1790. Valuable for its late appearing fls. See P. & G. 6: 56–58. 1950.

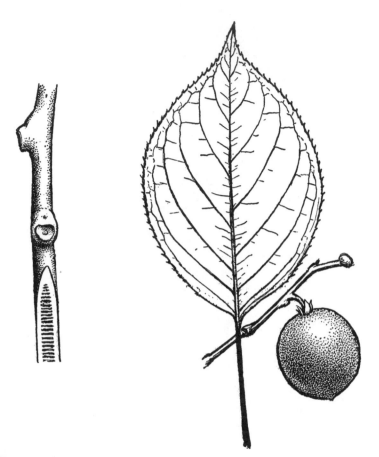

FIG. 86. *Actinidia arguta*, Bower Actinidia. Brt. × 5; lf. and fr. nat. size.

TAMARICÀCEAE — TAMARISK FAMILY

TÁMARIX — TAMARISK

The French Tamarisk, *T. gállica*, is one of the most popular spp. of this shrub or small tree. Lvs. scale-like, *deciduous;* brts. resemble superficially those of a *Thuja* or *Chamaecyparis*, hence sometimes, as in Bermuda, popularly known as "Cedar," but lvs. are *alternate*. Tiny pink fls. are borne in great profusion along brts. from June to Aug. Mediterr. region.

FLACOURTIACEAE — FLACOURTIA FAMILY

A family of mostly tropical trees and shrubs with usually alternate, long-petioled, leathery lvs.

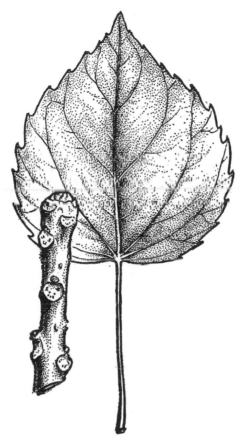

FIG. 86a. *Idesia polycarpa.* Lf. and tip of brt. showing true terminal bud, small lat. buds and lf. scars. Nat. size. (Lf. a little smaller than average.)

The Idesia, *I. polycarpa* Maxim., is a tree to 50 ft. with close, grayish white bark; ovate, acuminate, and cordate, long-petioled simple lvs. 3–10 in. long; true terminal bud present. Apparently hardy s. of New York City. Specimen at Brooklyn Botanic Garden died back a good deal during cold period of 1933–34. S. Japan; e. and w. China.

THYMELAEÀCEAE — MEZEREUM FAMILY

DÁPHNE

Low, deciduous or evergreen, exotic shrubs, popular as ornamentals, blooming in Mar. or Apr. Three spp. are commonly cult. which may be distinguished by the following key.

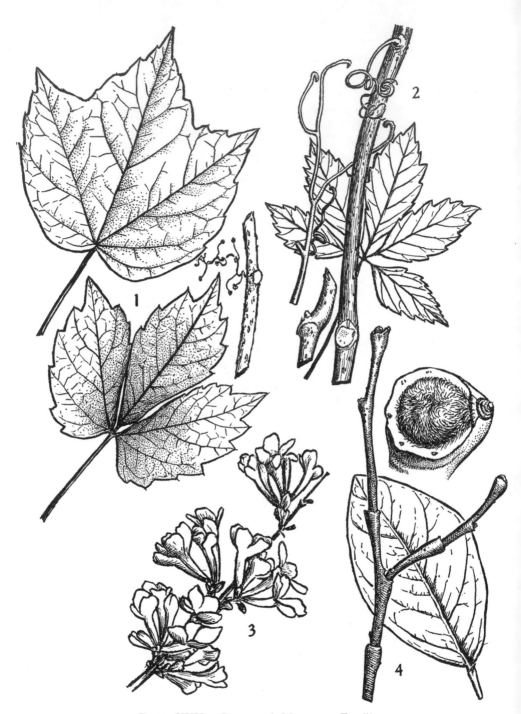

PLATE XXX. Grape and Mezereum Families:
Virginia Creeper, Boston-ivy, Leatherwood, and Daphne.

1. *Parthenocissus tricuspidata*, Boston-ivy. Two forms of lvs. and brt. with tendrils nat. size.
2. *P. quinquefolia*, Virginia Creeper, Woodbine. Lf., brts. and young tendrils nat. size. (Tendrils when mature usually show flattened discs at tip.)
3. *Daphne mezereum*, February D. Brt. with fls. nat. size.
4. *Dirca palustris*, Leatherwood. Lf. and brt. nat. size; topmost lat. bud showing leaf scar, bundle scars and end of season's growth × 6.

Lvs. deciduous.

  Lvs. opposite, silky beneath; fls. lilac..........................*D. genkwa*

  Lvs. alternate, glabrous; fls. rosy-purple to white (Pl. XXX, 3)..*D. mezereum*

Lvs. evergreen; low, procumbent shrub with pink fls.................*D. cneorum*

    The Lilac D., *D. génkwa* (Sieb. and Zucc.), has fls. in early Apr. much before lvs.
China, Korea. The Rose D. or Garland Flower, *D. cneòrum* L., with rosy pink fls.,
grows close to the ground and is the best of the evergreen D's. Sometimes languishes
in cult. Mts. of c. and s. Eu.

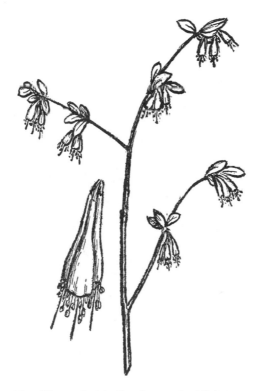

Fig. 87. *Dirca palustris*, Leatherwood. Fl. brt. nat. size.

Dírca

    [0]**D. palustris.** Leatherwood, Wicopy (Pl. XXX, 4 and fig. 87). Shrub to 6 ft.;
brts. very tough, swollen at nodes; lf. scars, with 5 bundle traces, almost enclosing
silky bud; no true terminal bud. Fls. yellow in very early spring. **Never very common.** N.B. to Fla. and w. Bark used by Indians to make thongs.

ELAEAGNÀCEAE — OLEASTER FAMILY

Shrubs or small trees with tiny silvery or sometimes also brown scales.

ELAEÁGNUS

Lvs. alternate

The Russian-olive, or Oleaster, **E. angustifòlia** L. (Fig. 88), to 20 ft. is unmistakable because of its silvery lvs. and brts.; the latter sometimes with thorns.   Lvs. oblong to linear-lanceolate, 1½–3 in. long.   Tiny silvery scales (no brown ones) cover brts., lvs. (especially beneath), fls., and fr. which is a silvery yellow drupe.   Often cult. and sometimes escaped.   S. Eu. and Asia.   Another sp., ⁰*E. umbellàta* Thunb. (Pl. XXIX, 3), **the**

FIG. 88. *Elaeagnus angustifolia*, Russian-olive, Oleaster.   Brt., lvs., and fr. **nat.** size; tip of brt. × 4.   Lvs. often narrower than in fig.

Autumn E., has more juicy fr., red-brown to pink, dotted with both brown and silvery scales, and lvs. often have a few brown scales beneath.   Lvs. elliptic to ovate-oblong, 1–3 in. long.   Cult.   China, Korea, Japan.

SHEPHÉRDIA

Lvs. opposite

The Soapberry, ⁰**S. canadénsis** (L.) Nutt., has lvs. nearly scale-less and green above; silvery downy and scurfy with scattered brown scales beneath.   A low shrub — to 6 ft. Nfd. to n. N.E. to n. and w. N.Y. and w.

The Silver Buffaloberry, *S. argentea* Nutt., with lvs. silvery on both sides, and with scales merely dentate, while in *S. canadensis* they are cleft to the base, is much planted in the midwest and northwest, and often tree-like.

NYSSÀCEAE — SOUR-GUM FAMILY

NÝSSA — TUPELO

N. sylvática Marsh.   Black Tupelo, Pepperidge, Black- or Sour-gum (Fig. 89).   Large tree, in moist soil; the many short, wide-angled

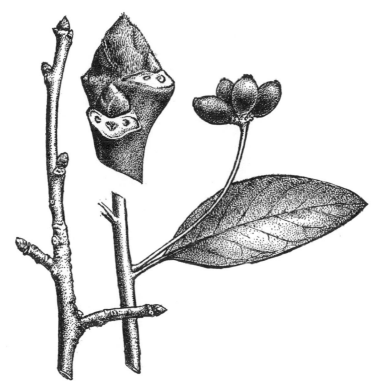

FIG. 89.   *Nyssa sylvatica*, Black Tupelo, Pepperidge.   Lf. fr., and brt. nat. size; terminal bud and last laterals × 6.

brs. reminding one of the pear or hawthorn.   Lvs. alternate, simple, narrowly obovate, turning crimson very early, sometimes in Aug.; *pith with unequally spaced diaphragms;* lf. scars reddish brown, with 3 *very distinct vascular bundle scars in a straight or slightly curved row;* stipule scars lacking; fr. small, a dark blue drupe; buds smooth, ovoid, pointed, dark red-brown; outer scales glabrous, inner scales (showing at tip) hairy; true terminal bud present, about 4 scales exposed.   Me. to Fla. and w.

DAVÍDIA

The Dove-tree, *D. involucràta* Baill. (Fig. 90), a striking tree when in fl., with its large, creamy-white floral bracts and large bright green lvs., is hardy only in the southern part of our area.   Pith diaphragmed.   Good specimens at the Brooklyn Bot. Gard. W. China.

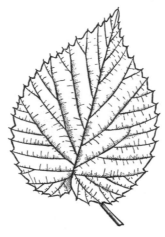

FIG. 90.   *Davidia involucrata*, Dove-tree.   Lf. × ½.

ARALIÀCEAE — GINSENG FAMILY

HÉDERA — IVY

The English Ivy, [0]*H. hélix* L., vies with the Japanese Creeper or Boston-ivy for first place as a wall cover, but most vars. are not quite as hardy as the latter.   However, it has the advantage of being evergreen. It climbs by aerial rootlets (not tendrils as in *Parthenocissus*).   The lvs. are of 2 kinds; those of vegetative brts. being 3–5-lobed, and those of flowering brts. being ovate to rhombic, tending to be without lobes (Fig. 91).   Cult. since ancient times.   Many garden forms and geographical vars. are known.   Eu.   See *Brooklyn Botanic Garden Lfts.* Ser. XXVIII No. 1–3. 1941.

ACANTHOPÁNAX

[0]*A. sieboldiànus* Mak. (Pl. XXIX, 4), a shrub to 9 ft. distinguished from other spp. by thorny brts. and by lvs. with 5–7 lfts. partly fascicled on short brts., is often cult. Japan.

FIG. 91. *Hedera helix*, English Ivy. Fr. brt. showing gradual change to lvs. without lobes; below, ordinary lf. × ½.

ARÀLIA

A. spinòsa L. Devils-walking-stick, Hercules'-club. (Fig. 92). A tree with prickly trunk and brts., to 45 ft.; lvs. very large, 1–3 ft. long, bipinnate and prickly; large clusters of whitish, small fls. in Aug., and black fr. in autumn; lf. scars half or more encircling stem, bundle traces about 5. Cult. and becoming natzd. in southern part of our area. S. Pa. to Fla. and w.

The Bristly A., ⁰**A. hìspida** Vent., is a native subshrub, to 3 ft. with partly woody stems and bipinnate lvs. occurring from Nfd. to w. N.C. and w.

CORNÀCEAE — DOGWOOD FAMILY

CÓRNUS — DOGWOOD

Lvs. opposite (alternate in **C. alternifòlia**), entire; fls. small, in rather close bunches (cymes or heads); lf. buds narrow and with short stalks, with a pair of nearly or quite valvate scales. Slicing off the bud lengthwise with a razor blade will determine the stalked character when one arrives at the center. Fr. a small drupe.

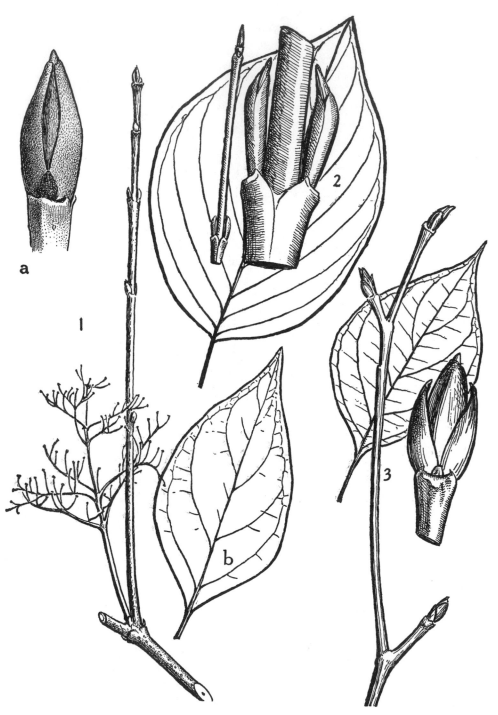

PLATE XXXI. Dogwoods.

1. *Cornus racemosa*, Gray D., Panicled D.   Lf. (b) and brt. nat. size, showing old fr. stalks;
    a, terminal bud complex × 5.
2. *C. rugosa*, Roundleaf D.   Lf. and brt. nat. size; part of brt. with lat. buds × 5.
3. *C. alternifolia*, Pagoda D., Alternate-leaf D.   Lf. and brt. nat. size; terminal bud complex × 5.

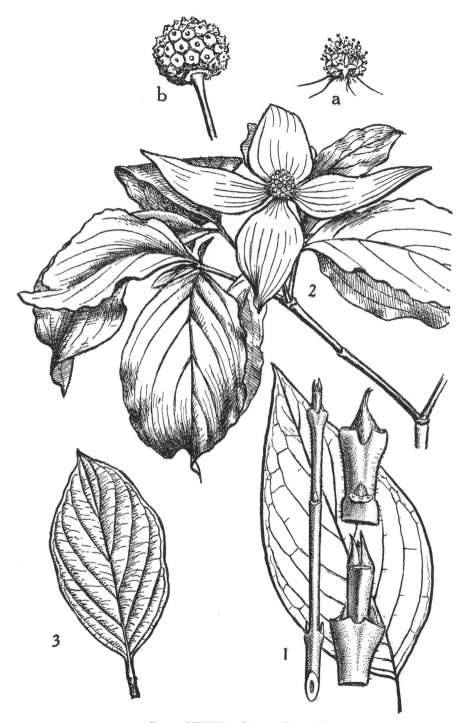

PLATE XXXII. Dogwoods concl.

1. *C. amomum*, Silky D.   Winter brt. and lf. nat. size; terminal bud complexes, 2 views, × 4.
2. *C. kousa*, Kousa D.   Brt. and lvs. with fl. head (unopened) and fl. brts. nat. size; a, head of opening fls.; b, head of fr., both about nat. size.
3. *C. alba*, Tatarian D.   Lf. nat. size.

FIG. 92. *Aralia spinosa*, Hercules'-club (Devils-walking-stick SPN). Lf. × ⅙.

KEY TO DOGWOODS

Lvs. (and buds and brts.) alternate (Pl. XXXI, 3)...............7. *C. alternifolia* 195
Lvs. (and buds and brts.) opposite.
  Fl. buds much larger than lf. buds, or of different shape; small trees.
    Outer fl. bud scales expanding at fl. time to form 4 large white bracts, usually
      notched at tip (Fig. 93), native...........................1. *C. florida* 192
    Outer fl. bud scales not, or little expanding at fl. time, or deciduous; cult.
      Head of fls. subtended by 4 small, yellowish bracts.............9. *C. mas* 195
      Head of fls. subtended by 4 large, white, pointed bracts (Pl. XXXII, 2)
                                          2. *C. kousa* 193
  No marked difference between fl. and lf. buds; shrubs.
    Fl. clusters elongated (old fr. stalks persistent in fall and winter (Pl. XXXI, 1)
                                     6. *C. racemosa* 194
    Fl. clusters more or less flat-topped.
      Pith of 2-yr.-old brts. brown or tawny; fr. light blue; brts. downy at tip with
        rusty hairs (Pl. XXXII, 1)..........................3. *C. amomum* 193
      Pith of 2-yr.-old brts. white; fr. white.
        Brts. bright yellow; cult......under no. 4. *C. stolonifera* var. *flaviramea* 194
        Brts. green, spotted with purple; lvs. comparatively large, more or less
          rounded, woolly beneath; native (Pl. XXXI, 2)........8. *C. rugosa* 195
        Brts. some shade of red.
          Brts. scarlet or carmine; stoloniferous; native (Fig. 94) 4. *C. stolonifera* 193
          Brts. blood red; not stoloniferous; cult. (Pl. XXXII, 3)....5. *C. alba* 194

1. C. flòrida L. Flowering Dogwood (Fig. 93). Small tree; bark appearing like alligator skin; lvs. pale beneath; fr. a bright red drupe; fl. buds globular, lf. buds narrow. The only native sp. of this region in which the bud scales of the fl. buds grow out into large petal-like bracts at the time of flowering. In the Pacific coast region *C. núttalli*, Pacific D., is similar. Very common. One of our handsomest small trees when in fl. Often cult. Pink and creamy-yellow, wild as well as cult., vars. are often seen. Sw. Me. to N.Y. and Ont. to Fla. and w.

FIG. 93. *Cornus florida*, Flowering Dogwood. Brt. with fl. and lf. buds; fr. and lf. × ½.

2. In the Kousa D. *Cornus kousa* Hance (Pl. XXXII, 2), the outer bud scales are deciduous at flowering time. The white petal-like, pointed bracts develop from *inner* scales, and the small frs. grow together into a tight, fleshy head. Bark thin and flaky, exfoliating. A handsome small tree. Japan, Korea. Sometimes cult. Doubtfully hardy n. of Mass.

3. ⁰C. amòmum Mill. Silky D., Kinnikinnick (Pl. XXXII, 1). Lvs. silky, downy, and often rusty beneath; brs. red, *pith tawny* (this shows best in stems 3 yrs. old or more); fr. pale blue. Moist or wet ground; very common. S. Me. to Ga. and w.

4. ⁰C. stolonífera Michx. Red-osier D. (Fig. 94). Brs. red, with *white pith* at all ages; lvs. whitish beneath, pubescent on both surfaces; *fr. white or lead color;* stems (stolons) when they touch the soil root easily, making new plants. Wet places; distinguished from last by stolons, white fr., and *white pith;* also brts. have a brighter scarlet hue.

Nfd. to D.C. and **w.**   Var. *flavirámea* (Spaeth) Rehd., Yellow-twig D., has bright yellow brts.   Often cult.

5. The Tatarian D., *Cornus alba* L. (Pl. XXXII, 3), strongly resembles **C. amo-mum,** but the brts. are more scarlet or carmine, like those of **C. stolonifera**; pith white; fr. white or slightly bluish; lvs. are usually only acute, while in **C. stolonifera** they are usually more acuminate.   Cult. and much used in landscape planting.   Asia.

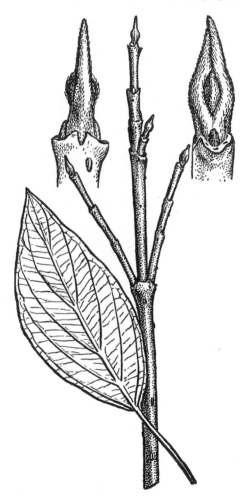

Fɪɢ. 94.   *Cornus stolonifera*, Red Osier Dogwood.   Lf. and brts. nat. size; 2 views of terminal bud complex × 5.

6. ⁰**C. racemòsa** Lam.   Gray D., Panicled D. (Pl. **XXXI**, 1).   Common, and often in drier soil, along fences, etc.   Fls. in somewhat convex or elongated clusters (paniculate), not flat-topped as in 3 preceding spp.;

brts. gray or reddish gray; pith brown in brts. of 2 yrs. or more, white in 1 yr. brts;. fr. white, on red stalks; lvs. glaucous below. Me. to Md. and w.

7. **C. alternifòlia** L.f. Alternate-leaf D., Pagoda D. (Pl. XXXI, 3). Small tree; lvs. alternate, pale beneath; brts. arranged usually in tiers; pith white; fr. blue-black, glaucous. Nfd. to Ga. and w.

8. °**C. rugòsa** Lam. (*C. circinàta* L'Her.). Roundleaf D. (Pl. XXXI, 2). Lvs. large, *almost circular, woolly beneath;* brts. green or pinkish; pith white; fr. light blue. Unmistakable from the shape and size of the lvs. and the peculiar hues of the brts. Not common, but to be looked for on rocky slopes. Que. to Va. and w.

9. The Cornelian-cherry D., *C. mas* L., a small tree, has *yellow fls.* blooming early in Apr., enclosed, in the winter-bud stage, by 4 scales which, however, do not expand at flowering time as in **C. florida.** Commonly cult. Eurasia.

ERICÀCEAE — HEATH FAMILY

A large family, including huckleberries, blueberries, cranberries, mountain-laurel, rhododendron, azalea, trailing-arbutus, etc. In most of the genera the anthers open by pores at the tip. Lvs. simple, often evergreen; corolla usually with united petals (petals distinct in *Clethra, Chimaphila, Ledum,* and *Leiophyllum*).

KEY TO GENERA AND SOME SPECIES

This key does not include *Chimaphila, Ledum, Leiophyllum, Menziesia, Phyllodoce,* or *Andromeda,* which see.

A. Lvs. evergreen.
  B. Lvs. small, not more than ⅖ in. long, needle-like or scale-like.
    C. Lvs. needle-like, in whorls of 3 or 4 (Fig. 100, p. 209).............*Erica* 207
    C. Lvs. scale-like, opposite, in 4 rows lengthwise on stem............*Calluna* 207
  B. Lvs. larger, but averaging less than 1 in. long, entire.
    C. Fr. a berry (containing many small seeds).
      D. Berries white, aromatic; small trailing subshrub of bogs or mossy woods; northern, or s. in uplands...........*Gaultheria* (*Chiogenes*) *hispidula* 207
      D. Berries some shade of red; small prostrate shrubs; lvs. oval or elliptical, obovate.
        E. Plants of dry, rocky, or peaty acid soil; northern (Pl. XXXVIII, 5, a, b)........................ *Vaccinium vitis-idaea,* var. *minus* 212
        E. Plants of bogs, not limited to northern regions or uplands (Pl. XXXVII, 5)........................... *Vaccinium oxycoccus* and *macrocarpon* 212
    C. Fr. a drupe with a single stone (composed of 5–10 wholly or partly fused nutlets), dull red, dry and mealy; lvs. obovate, blunt, or rounded at tip (Pl. XXXVII, 6)...*Arctostaphylos* 207
  B. Lvs. averaging more than 1 in. long.
    C. Fr. a berry, composed in part of the fleshy calyx; with wintergreen flavor; lvs. serrate (Pl. XXXV, 4).......................*Gaultheria procumbens* 207

PLATE XXXIII. Heath Family:
Rhodora, Swamp-honeysuckle, Sweet Pepperbush, and Sand-myrtle.

1. *Rhododendron canadense*, Rhodora. Fruiting brt. nat. size; terminal (fl.) bud complex × 5.
2. *R. viscosum*, Swamp-honeysuckle. Lf. and winter brt. nat. size; terminal bud complex (fl. and lf. buds) × 3.
3. *Clethra alnifolia*, Sweet Pepperbush. Fruiting brt. and lf. nat. size; terminal bud and stem tip × 5; fr. × 2.
4. *Leiophyllum buxifolium*, Sand-myrtle. a, fl. cluster; b, fr. cluster nat. size.

Cléthra — White-alder

⁰**C. alnifòlia** L.   Sweet Pepperbush (Pl. XXXIII, 3).   Shrub with long, erect spikes of white, fragrant fls. in July or Aug. and dry frs.; buds minutely pubescent; terminal buds large and pointed, lateral buds very small and inconspicuous or developing into short brts. the same season; brts. downy or scurfy, angled; outer bark peeling off the second yr.; lf. scar oval or triangular, with bundle scar forming a prominent semicircular or broad u-shaped ridge; lvs. obovate, pointed, smooth, serrate.   In wet ground or swamps; also cult. s. Me. to Fla. and w.

Chimáphila

The Prince's-pine or Pipsissewa, ⁰C. **umbellàta**, (L.) Bart. var. **cisatlántica** Blake and the Spotted Wintergreen, ⁰C. **maculàta**, (L.) Pursh., are common in rich woods in general all through our area.   Low, somewhat woody plants, evergreen, usually not more than 6 or 8 in. high, with exquisite white or pinkish waxy fls. in summer.   In the first sp. the lvs. are bright green; in the second, they are variegated with white along the veins.

Lèdum

The Labrador Tea, ⁰L. **groenlándicum** Oed., is found usually in bogs, from Greenland to Pa. and w.   Low shrubs with entire, evergreen lvs., rusty woolly beneath.

Rhododéndron

By some authorities divided into 2 genera; *Azàlea*, with deciduous lvs., and *Rhododéndron* with typically persistent, i.e. evergreen lvs.   The cult. spp., vars., and forms of both *Azalea* and *Rhododendron* are legion, many of them results of plant breeding, others natives of Japan, China, India, etc.   For furnishing masses of brilliant color in landscape design they are unequaled.   They thrive best in acid soil, for which a continual mulch of dead (oak) lvs. is helpful.

Of the cult. forms, ⁰R. *mucronulàtum* Turcz., Korean R., from Asia, with rose purple fls. and deciduous lvs., is one of the earliest sp., flowering in Mar. and Apr. in N.Y. and N.E.   Another handsome, rather early sp. (May and June) not very showy but dainty and exquisitely beautiful, is *⁰R. carolinianum* Rehd., Carolina R., a native of N.C., a "true" R., with rose-purple fls., and lvs. which are rusty beneath due to a covering of tiny scales.   For other cult. forms see Rehder's Manual (49), pp. 694–723.

R. máximum L.   Rosebay R., Great-laurel, to 30 ft., with thick, evergreen lvs., and pale rose to nearly white fls., is found rarely in deep, damp woods from Me. to Ga. and w. and is closely related to many cult. forms which bloom in June and July.   More common southward.

The Catawba R., *R. catawbiense*, occurs naturally and is abundant in the Southern Appalachians from Va. to n. Ala.—one of our most beautiful native shrubs.   The fls. are lilac-purple and the lvs. are glabrous beneath.   *R. maximum* is a similar sp. occurring also farther n., the fls. being rose-colored or white and the lvs. pubescent beneath (rarely nearly glabrous).

⁰**R. canadénse** (L.) Torr. Rhodora (Pl. XXXIII, 1), belongs in the Azalea group of deciduous-lvd. R.   A low shrub, about 3 ft. in height and distinguished by its pale more or less pubescent lvs. and pale to deep rose-purple fls.; stamens 10, petals nearly or quite distinct to the base. Swamps and damp slopes.   Nfd. to n. N.J. and ne. Pa.

The Flame- or Yellow-Azalea, ⁰**R. calendulàceum** (Michx.) Torr., has hairy lvs. and large, orange blossoms, turning to flame-color, and without fragrance.   Woods.   Comes into our area in sw. Pa.   Often cult.

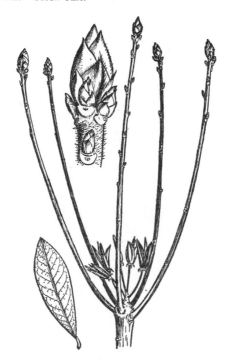

FIG. 95.   *Rhododendron nudiflorum,* Pinxter-flower.   Lf. and brt. showing lf. and fl. buds × ½; above, terminal fl. bud complex × 1½.

⁰**R. viscòsum** (L.) Torr.   Swamp-honeysuckle, Clammy Azalea (Pl. XXXIII, 2).   Lvs. deciduous, shining, somewhat narrowed toward their bases, often arranged in a conspicuous, flat mosaic near tip of brt.; *brts. bristly;* fl. buds large, with ciliate scales, terminal lf. buds much smaller; fls. *after* the lvs. in June and July, white, fragrant, sticky.   *Grows in swamps or moist soil.*   Sw. Me. to S.C. and w.

⁰**R. nudiflòrum** (L.) Torr. (Fig. 95).   Pinxter-flower, Purple-Honey-suckle.   Much like the last, but usually not so tall; with duller lvs. and generally *smoother* brts.; fls. opening much earlier, *with the unfolding of the lvs.;* usually of various shades of pink; buds as in the last.   *Grows in*

drier soil. From the winter characters alone, this and the last sp. are difficult to distinguish. Mass. to S.C. and w.

MENZIÈSIA

The Allegany Menziesia, or Minnie-bush, ⁰**Menzièsia pilòsa** (Michx.) Juss., a low shrub with small, bell-shaped, yellowish-white or pinkish fls., deciduous, elliptic to oblong-obovate lvs., comes into our area in Pa. Sometimes planted in rock gardens.

KÁLMIA

Smooth shrubs with showy fls., evergreen lvs., and small buds with only 2 scales showing.

K. latifòlia L. Mountain-laurel (Fig. 96). Has ovate, thick, leathery, *alternate* lvs. Fls. terminal. N.B. to Fla. and w.

FIG. 96. *Kalmia latifolia*, Mountain-Laurel. Brt. with fl. buds and lvs. × ½.

⁰**K. angustifòlia** L. Lambkill, Sheep-l. (Pl. XXXV, 3). A small shrub with smaller, thinner, oblong lvs., which are *opposite* or in 3's (rarely in 4's); fls. lateral. Poisonous to stock. Lab. to Ga. and w.

The Pale or Bog-laurel, ⁰**K. polifòlia** Wang., with 2-edged brts. and small, opposite lvs. (rarely in 3's), lanceolate or linear and white beneath, and pink or crimson fls., ½ in. or more in diam., is found only in bogs. Nfd. to n. N.J., Pa. and w.

PHYLLÓDOCE

The Mountain-heath, ⁰**P. caerùlea** (L.) Bab., with evergreen, alternate, linear lvs. about ¼ in. long, and small, cylindric or bell-shaped purplish fls., a low shrub to 6 in., native in the far north, comes into our area in Me. and N.H.

ANDRÓMEDA

The Bog-rosemary, ⁰**A. glaucophýlla** Link, with evergreen, linear lvs., glaucous beneath, their margins inrolled, and umbels of small pink or white round-urn-shaped fls. is sometimes found in bogs from Lab. to n. N.J., Pa. and w. Sometimes planted in rock gardens in the colder parts of our area.

ENKIÁNTHUS

The Redvein E., *E. campanulàtus* (Miq.) Nichols (Pl. XXXIV, 1), and White E., ⁰*E. perulàtus* (Miq.) Schneid. (Pl. XXXIV, 2), both natives of Japan, are popular ornamental shrubs. The former is the more attractive, having larger fls. of a slightly yellowish hue, veined with red. The lvs. of this sp. turn brilliant red in the fall. In both spp. the lvs. are crowded together at the ends of the branches, and when not in flower the plants are easily recognized by this bunched or whorl-like appearance of the lvs. along with a general resemblance to other members of the family, especially to *Pieris*. Lvs. deciduous.

PÍERIS

The Mountain P., *⁰*P. floribúnda* (Pursh) Benth. and Hook. (Pl. XXXV, 2) and Japanese P., *P. japònica* (Thunb.) D. Don. (Pl. XXXV, 1), are best distinguished by the position of their fl. clusters; more or less erect in the former and always drooping in the latter. *P. japonica* is the more popular in landscape planting. Also known respectively in the trade as Mt. and Japanese Andromeda.

LEIOPHÝLLUM — SAND-MYRTLE

The Sand-myrtle, ⁰Leiophyllum buxifòlium (Berg.) Ell. (Pl. XXXIII, 4), is a small shrub 4 in. to 3 ft. high with small oval or oblong, shining, evergreen lvs. up to ½ in. long, and with small white fls. in terminal umbel-clusters. Sandy pine barrens, N.J.

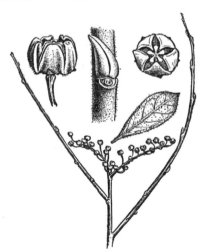

FIG. 97. *Lyonia ligustrina*, Maleberry. Fr., brt., and lf. × ½; above, 2 views of fr. × 2½; lat. bud × 2½.

LYÒNIA

⁰**L. mariàna** (L.) D. Don. Stagger-bush (Pl. XXXVI, 1 and 2). Buds small, roundish, crimson, standing out at a wide angle from the smooth, yellow brts., and urn-shaped, persistent dry frs.; buds with at least 4 scales exposed; true terminal bud lacking. Poisonous to stock. S.N.E. and se. N.Y. to N.J., Pa. and w. Fls. handsome, white or pinkish, of the "Blueberry" type.

PLATE XXXIV. Heath Family cont.
Enkianthus, Leather-leaf, and Trailing-arbutus.

1. *Enkianthus campanulatus*, Redvein E. Flowering brt. nat. size.
2. *E. perulatus*, White E. Flowering brt. nat. size.
3. *Chamaedaphne calyculata*, var. angustifolia, Leather-leaf. Brt. with fl. buds nat. size; fl. bud
   × about 10.
4. *Epigaea repens*, Trailing-arbutus. Brt. with fl. buds nat. size.

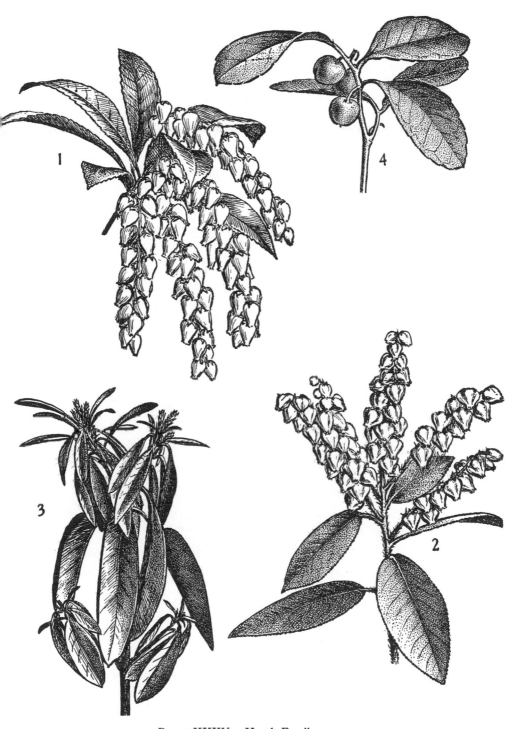

PLATE XXXV. Heath Family cont.
Pieris, Sheep-laurel, and Wintergreen.

1. *Pieris japonica*, Japanese Pieris. Flowering brt. nat. size.
2. *P. floribunda*, Mountain P. Flowering brt. nat. size.
3. *Kalmia angustifolia*, Sheep-laurel. Brt. showing lvs. and fl. buds nat. size.
4. *Gaultheria procumbens*, Aromatic Wintergreen. Brt. with fr. nat. size.

⁰**L. ligustrìna** (L.) DC.   Maleberry.   (Fig. 97).   Has one-sided racemes of small, white fls., and roundish, dry frs.; brts. often minutely pubescent, *yellow;* buds smooth, slender, sharp-pointed, *crimson*, with 2 scales showing, flattened, lying close to brts.; true terminal bud lacking; lf. scars shield-shaped; lvs. obovate, oblong or elliptic, pointed, entire or minutely serrulate.   N.E. to Ga. and w.

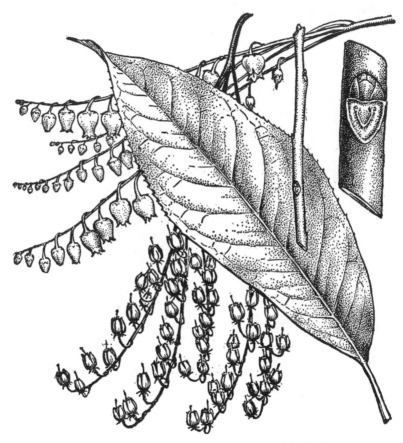

Fig. 98.   *Oxydendrum arboreum,* Sourwood.   Lf., fls., brt., and fr. nat. size; brt. and bud × 3.

CHAMAEDÁPHNE

⁰**C. calyculàta** (L.) Moench var. **angustifòlia** (Ait.) Rehd.   Leatherleaf (Pl. XXXIV, 3).   Low, much branched shrub to 3 ft., of bogs and pond margins.   Lvs. nearly evergreen, scurfy beneath; fls. in one-sided leafy racemes.   Nfd. to Ga. and w.

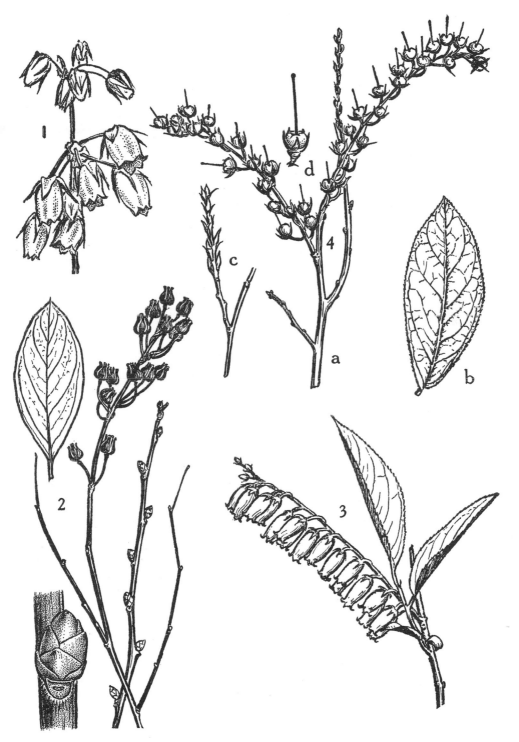

PLATE XXXVI. Heath Family cont. Lyonia and Leucothoe. 1. *Lyonia mariana*, Stagger-bush. Fl. cluster nat. size. 2. Same, brt. with fr., showing urn-shaped capsules; brts. with lf. and fl. buds all nat. size; fl. bud and part of brt. × 5. 3. *Leucothoe racemosa*, Fetter-bush. Flowering brt. with lvs. nat. size. 4. Same, a, fruiting brts., showing globular capsules; b, lf.; c, fl. buds—all nat. size; d, fr. × 2 (showing persistent style).

Oxydéndrum

O. arbòreum (L.) DC. Sourwood, Sorrel-tree (Fig. 98). Tree, sometimes 50–60 ft. tall, with deeply furrowed bark, no true terminal bud; axillary buds minute, sunken in bark, obtuse, with about 4 outer dark red scales; lvs. simple, serrulate and acuminate, with an acid taste, hence the name; fls. in one-sided racemes in late summer; drooping clusters of dry fr. conspicuous all through fall and winter. Lvs. turn scarlet in fall. Becoming increasingly popular for ornamental planting in n. where it is hardy to eastern Mass. N.J. and Pa. w. and s. Good planted specimens along Hutchinson R. Pkwy. in Westchester Co., N.Y.

Leucóthoë

⁰**L. racemòsa** (L.) Gray. Leucothoë, Fetter-bush, Sweet-bells (Pl. XXXVI, 3, 4). Shrub with thin, deciduous, alternate, toothed lvs. and very fragrant, white fls. in one-sided, terminal racemes; in winter the

FIG. 99. *Leucothoe editorum*, Drooping Leucothoe. Brt. with fl. buds nat. size.

roundish dry frs. are still to be seen as well as the racemes of fl. buds for the next yr.; brts. often reddish-brown above, green below; buds small, roundish; lf. scars crescent-shaped or semicircular, with one central bundle scar; true terminal bud lacking; poisonous to stock. S.N.E. and se. N.Y. to Fla. to La. [*Eubotrys racemosa* (L.) Nutt. See (16)].

The Drooping L., *[0]*editorum* Fern. and Schub. (Fig. 99) has evergreen lvs. up to 6 in. long, long-pointed, sharply serrulate; fl. buds naked through winter. A low shrub in s. Appalachians, much used in landscape work. Mts., Va. to Ga. and Tenn.

Epigaèa

The Trailing-arbutus, [0]**E. rèpens** L. (Pl. XXXIV, 4) is a small, prostrate shrub with creeping stems; evergreen, hairy, oval or rounded lvs., and fragrant rose-colored or white fls. in very early spring. Also known popularly as Mayflower, but unfortunately this name has been applied to many other early spring fls. Sandy to peaty woods or clearings. S.N.E. and se. N.Y. to Fla. and w.

Gaulthèria

The Aromatic Wintergreen, Teaberry, or Checkerberry, [0]**G. procúmbens** L. (Pl. XXXV, 4), has slender stems, creeping on the ground or just below; lvs. evergreen with short petioles, oval or obovate, obscurely serrate, smooth, and shining; lvs. and bright red "berries" with characteristic "wintergreen" flavor. Nfd. to Ga. and w.

The Creeping Snowberry, [0]*Chiogenes hispidula* (L.) T. & G. is recognized by Fernald in Gray's Manual (8th ed.) as [0]**Gaultheria hispídula** (L.) Bigel., and is a small, trailing subshrub with bright white, juicy and aromatic berries and small lvs., bristly beneath. Bogs and mossy woods; Lab. to Pa., upland to N.C. and w.

Arctostáphylos

The Bearberry, [0]**A. ùva-úrsi** (L.) Spreng. (Pl. XXXVII, 6), also known as Kinnikinnick (see p. 193), Mealberry, Hog-cranberry, is a trailing shrub, with exfoliating bark and small, rounded, thick, evergreen lvs. In our area represented by the var. coáctilis Fern. and Macbr., in which the young brts. are permanently white-tomentulose. Nfd. to Va. and w.

Erìca — Heath

Heath and Heather are easily confused. In Heath the lvs. are acicular, i. e. short and stiff and needle-like, standing out from the brts., while in Heather they are smaller, scale-like, and imbricated on the brts. In the fls. of Heath the calyx is shorter than the corolla, while in Heather it is longer.

In the Spring Heath, [0]*Erica cárnea* L. (Fig. 100) the lvs., usually in whorls of 4, are glabrous and fls. are rose-colored, appearing very early, sometimes in Jan. or even before, during mild spells, but always by Mar. C. and s. Eu. Often cult. [0]*E. tétralix* L., the Cross-lvd. Heath of Eu., is sometimes natzd. in N.E. and N.J.

Callùna — Heather

For characters see above, under *Erica*.

The Scotch Heather, [0]**C. vulgàris** (L.) Hull, is sometimes natzd. in peaty or damp spots. Nfd. to N.J., mts. of W.Va., and w. Small, low, evergreen shrub, valued as an ornamental for its late fls. and often planted in large masses on sandy banks and slopes. There are many vars. of this, the only sp. Eu. and Asia Minor.

Gaylussàcia — Huckleberry

Fr. sweet, with *ten large* seeds*; lvs. (in our spp.) more or less dotted on the under surface with resin globules; true terminal bud lacking.

---

* Strictly, these are more than "seeds," being classed botanically as nutlets.

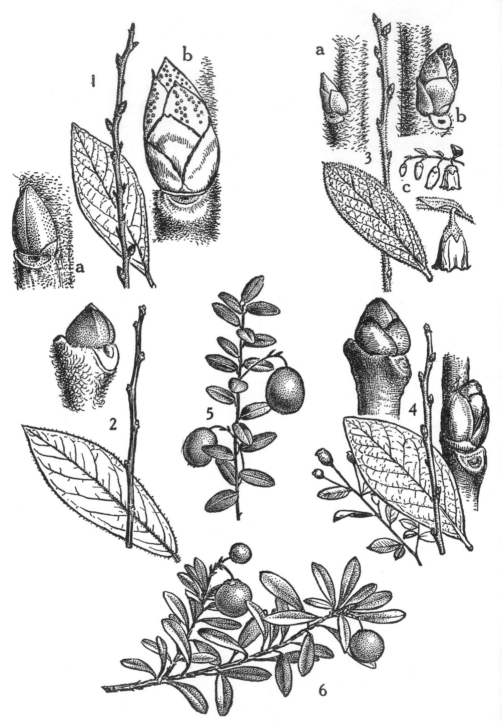

PLATE XXXVII. Heath Family cont. Huckleberries, Deerberry, Cranberry, and Bearberry. *Gaylussacia baccata*, Black Huckleberry. Brt. with lf. and fl. buds, and lf. nat. size; a, lf. bud × 1⋅ b, fl. bud showing resin globules × 10. 2. *G. frondosa*, Dangleberry. Brt. and lf. nat. size; abov⋅ apex of brt. showing last lat. bud, lf. scar and stub of end of season's growth (at left) × 10. 3. *⋅ dumosa*, Dwarf H. Brt. and lf. nat. size; a, lf. bud × 5; b, fl. bud showing resin globules × 5; c, cluster nat. size; and below, single fl. × 2. 4. *Vaccinium stamineum*, Deerberry. Brt. and lf. na⋅ size; rt., lat. bud × 10; above, apex of brt. showing last lat. bud, lf. scar and, at left, stub of end season's growth × 10; below, young fruiting brt. nat. size. 5. *Vaccinium macrocarpon*, Large or Ame⋅ Cranberry. Fruiting brt. nat. size. 6. *Arctostaphylos uva-ursi*, Bearberry. Fruiting brt. nat. size.

1. ⁰**G. dumòsa** (Andr.) T. and G. Dwarf Huckleberry (Pl. XXXVII, 3). Brts. arising from a horizontal, subterranean stem; lvs. oblanceolate, mucronate, glandular beneath; fls. bell-shaped, in racemes, white or pink, attractive; fr. black and usually tasteless; fl. buds much larger than lf. buds, the former sprinkled with resin dots. Rather rare or local. Se. N.Y. and e. Pa. to Fla. and w.

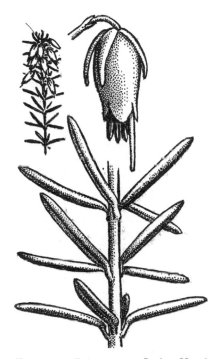

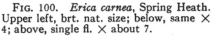

FIG. 100. *Erica carnea*, Spring Heath. Upper left, brt. nat. size; below, same × 4; above, single fl. × about 7.

FIG. 101. *Gaylussacia baccata*, Black Huckleberry. Brt. with fls.; fr. and cross section of fr. showing 10 seeds, all nat. size.

2. ⁰**G. frondòsa** (L.) T. and G. Dangleberry (Pl. XXXVII, 2). Stems slender, to 6 ft., smooth or somewhat pubescent; lvs. oval to obovate, *glaucous*, resin-dotted, and finely pubescent beneath; frs. dark blue, edible, long-stalked, with a pale bloom. Occasional. S.N.H., Mass. and se. N.Y. to Fla. and w.

3. ⁰**G. baccàta** (Wang.) K. Koch. Black Huckleberry (Pl. XXXVII, 1 and Fig. 101). Stems stiff, much branched, finely pubescent on younger parts; lvs. densely covered on under side with shiny resin globules which are sticky when young; fl. buds larger than lf. buds, sprinkled with resin globules; fr. black, shiny, edible, in short, dense

clusters; fls. tubular, more slender than those of blueberries, and more reddish. The common Huckleberry. Nfd. to Ga. and w.

Vaccìnium — Blueberry, Cranberry

Fr. usually more acid than that of huckleberries, and with *numerous small* seeds; true terminal bud lacking; fls. cylindrical or bell-shaped.

1. ⁰**V. stamíneum** L.　Deerberry (Pl. XXXVII, 4).　Low and much branched; brts. buff and usually downy; lvs. pale, glaucous and pubescent beneath, entire; fls. greenish-white or purplish, bell-shaped, with

Fɪɢ. 102.　*Vaccinium corymbosum*, Highbush Blueberry.　Fl. brt. nat. size.

projecting stamens; fr. greenish, with or without bloom, often pear-shaped, tart, on slender stalks about ½ in. long, resembling a green or greenish-yellow blueberry; fruiting brts. with small bracts.　Rather rare or local.　Mass. and N.Y. to Fla. and w.

2. ⁰**V. angustifòlium** Ait. var. **laevifòlium** House.　Low Sweet or Late Sweet Blueberry (Pl. XXXVIII, 3).　Low; stems smooth; lvs. *smooth*, shining, *green on both sides*, not mucronate; closely spinulose-serrulate; fr. mostly bluish black and glaucous, sweet. (*V. pensilvanicum* Lam.)　Common.　Nfd. to Va. and w.　Var. **nìgrum** (Wood) Dole, a black var. without bloom occurs.　*V. angustifolium* is lower and has smaller lvs. than var. *laevifolium*, and is often the commoner form.

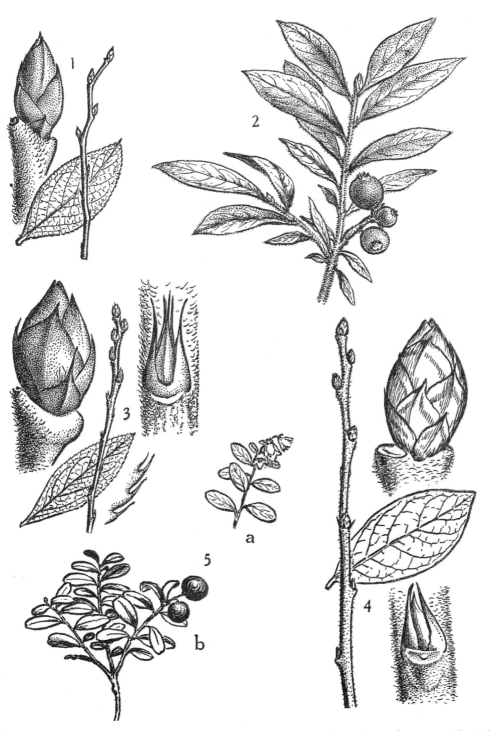

PLATE XXXVIII. Heath Family concl. Blueberries and Mountain-cranberry. 1. *Vaccinium vacillans*, Low Blueberry. Brt. and lf. nat. size; apex of stem, with uppermost lat. (fl.) bud and stub end of season's growth at left × 10. 2. *V. myrtilloides*, Canada B. Fruiting brt. nat. size. 3. *V. angustifolium*, var. *laevifolium*, Low Sweet B. Brt. and lf. nat. size; left, above, fl. bud × 10; rt., lf. bud × 10; below, rt., spinulose lf. serrations, enlarged. 4. *V. corymbosum*, Highbush B. Lf. and brt., latter showing fl. buds and two lf. buds nat. size; above, apex of stem, with uppermost lat. (fl.) bud, lf. scar (left), and stub of end of season's growth (rt.) × 10; below, lf. bud × 10. 5. *V. Vitis-idaea*, var. *minus*, Mountain-cranberry. a, Flowering brt., b, fruiting brt., both nat. size.

3. ⁰V. vacíllans Torr. Low Blueberry (Pl. XXXVIII, 1). Low; stems smooth and yellowish-green; lvs. dull above, glaucous below, *mucronate;* fr. blue, glaucous, very sweet. Common. Dry soil. N.E. and N.Y. to Ga. and w.

4. ⁰V. myrtilloìdes Michx. Canada or Velvet-Leaf B. (Pl. XXXVIII, 2). Pubescence marked in lvs. and stems, especially on young stems; berries rather large, more acid than in last two spp. Common, especially northward. Lab. to Va. and w. (*V. canadénse* Richards.)

5. ⁰V. corymbòsum L. Highbush B. (Pl. XXXVIII, 4 and fig. 102). Tall; brts. green or often reddish; fl. buds red, plump, pointed; lf. buds (smaller than fl. buds) pointed, the scales each with a prominent, spine-like point; lf. scars very narrow, sometimes merely transverse lines on brt.; lvs. entire, usually somewhat pubescent below; fr. blue-black, glaucous, sweet and juicy. Very common. Swamps or moist soil, but sometimes dry uplands. N.S. to Fla. and w.

6. ⁰V. atrococcum (Gray) Heller. Black Highbush B. Somewhat similar to the last, but has lvs. densely pubescent below, and *black, shiny fr.;* blooms and frs. earlier. N.E. to Fla. and w.

7. ⁰V. vìtis-idaèa L. var. mìnus Lodd. Mountain-cranberry, Cowberry (Pl. XXXVIII, 5, a, b). Small, creeping, evergreen shrub with thick, small, ovate, or obovate shining lvs.; berries red, acid, slightly bitter. Locally common along seacoast and mts. inland. Subarctic Amer. to Nfd. and n. N.E. and w.

8. ⁰V. macrocárpon Ait. Large or Amer. Cranberry (Pl. XXXVII, 5). Lvs. evergreen, oblong-elliptic; berry ⅖–⅘ in. in diam.; creeping, with fruiting brts. ascending. Bogs and swamps. Nfd. to N.C. and w. Much cult. for the market.

The Small Cranberry, 9. ⁰V. oxycóccos L., a native of Eurasia, also occurs over about the same range, the berries being ⅕ to nearly ⅖ in. in diam.

EBENÀCEAE — EBONY FAMILY

Diospỳros — Persimmon

D. virginiàna L. Common Persimmon (Fig. 103). Tree with alternate, ovate-oblong, entire, smooth lvs.; yellowish, fleshy frs. 1 in. or slightly less in diam., very puckery; somewhat triangular buds with only 2 or 3 greatly overlapping scales showing; bark deeply cut into thick, squarish segments; true terminal bud and stipule scars lacking; pith with irregular diaphragms, sometimes chambered. S. N.E. and se. N.Y. to Fla. and w.

The Kaki P., *D. kàki* L. f., a native of China and Japan, and often seen in our markets (from Calif.) has orange to bright yellow, edible frs., 1½–3 in. in diam. Not hardy in our area.

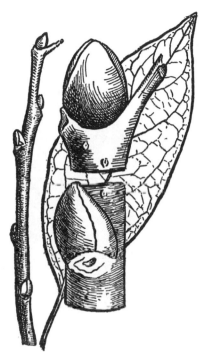

FIG. 103. *Diospyros virginiana*, Common Persimmon. Lft. and brt. nat. size; buds and stub showing end of year's growth × 4.

SYMPLOCÀCEAE — SWEETLEAF FAMILY

SÝMPLOCOS — SWEETLEAF

The Asiatic Sweetleaf, *Sýmplocos paniculàta* (Thunb.) Miq. (Fig. 103a), is sometimes cult. in our area. Lvs. simple, alternate, sharply serrulate, without stipules; fls. small, white, in panicles, 1½–3 in. long; fr. a *bright blue*, ellipsoid drupe about ¼ in. long; buds small, broadly conical, with 4–6 scales showing, solitary, or superposed; pith brown, not chambered. The native sp. is **S. tinctòria** L'Her., Common Sweetleaf, also known as Horse-Sugar from the eagerness with which cattle browse on the lvs., which are half-evergreen and obscurely toothed. In this sp. the fr. is orange or brown. Pith *chambered*. Del. to Fla. and w.

FIG. 103a. *Symplocos paniculata*, Asiatic Sweetleaf. Leafy brt. nat. size, and winter brt., enlarged. Note peculiar lobed (cat-face) lf. scars. These appear on long growths. See pp. 11, 12.

STYRACÀCEAE — STORAX FAMILY

STÝRAX — SNOWBELL

Buds small, stalked, *naked*, scurfy, superposed, terminal bud lacking; pith *green;* brts. rounded, slender, zigzag, rough scurfy. Two spp. are cult., the first more commonly.

S. japónica Sieb. and Zucc. Japanese Snowbell (Fig. 104). Small tree to 30 ft.; lvs. elliptic, 1–3 in. long, remotely toothed, these and the slender brts. with stellate pubescence when young; fls. white, pendulous in clusters of 3–6; fr. greenish, globular, pendent, with remains of the calyx prominent at its base.

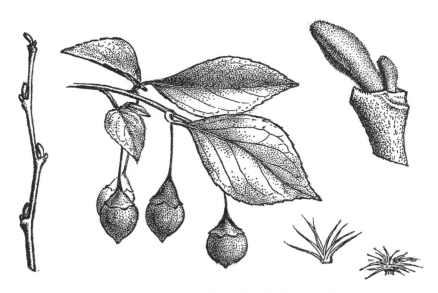

FIG. 104. *Styrax japonica*, Japanese Snowbell. Left, winter brt., nat. size; center, fr. brt. nat. size; right, topmost lat. bud superposed above small one, stub of stem end at left × 5; also, below, stellate hairs enlarged.

In the fragrant S., *S. obassia* Sieb. and Zucc., the lvs. are much larger (3–8 in. long), and proportionally broader, densely pubescent beneath; fls. fragrant, in racemes 4–8 in. long, often hidden by the large lvs. In this sp. the brts. are stouter and the bark exfoliates. Japan.

HALÈSIA — SILVERBELL

The Carolina Silverbell, *H. carolìna* L. (Fig. 105), native from Va. to Fla. and w., is frequently cult. in our area. Small tree to 30 ft. with *shreddy bark* even on young brts.; buds superposed, rather sharp-pointed, with about 4 fleshy red or greenish-red scales; pith rather small, *chambered*, white; fr. dry, with *4 prominent ridges or wings*; lvs. ovate, serrulate. Handsome especially in spring with its brts. lined with white, bell-shaped fls. The sp. with 2-winged fr., *H. diptera* Ell., is only rarely cult. and difficult to distinguish without the fr. (Old frs. or remains of them can usually be found on the ground.) S.C. to Fla. and w. *Halesia monticola* (Rehd.) Sarg. Mt. S., a larger sp., has been recommended as a street tree (45).

PTEROSTÝRAX — EPAULETTE-TREE

The Epaulette-tree, *P. hispida* Sieb. and Zucc. (Fig. 106), can be recognized by its buds, the terminal being elongated, hairy and naked, while the lateral are ovoid, mostly smooth and with 2 exposed scales. Like *Halesia* the twigs have quickly shredding bark; lvs. oblong to obovate-oblong, pointed, finely toothed. A rare tree in cult. Japan. Good specimens at N.Y. and Brooklyn Bot. Gardens.

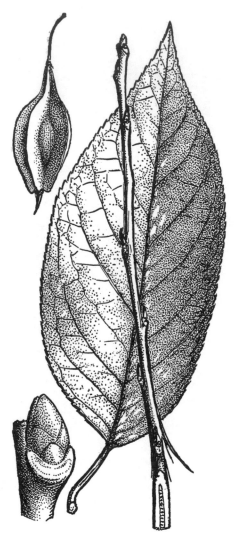

FIG. 105. *Halesia carolina*, Carolina Silverbell, Silverbell-tree. Lf., brt. (showing exfoliating bark and chambered pith) and fr. nat. size; lat. superposed buds × about 5.

OLEÀCEAE — OLIVE FAMILY

Trees or shrubs with opposite pinnate or simple lvs.; typically 4-parted fls. and 2 stamens; fr. a samara in the Ash, a capsule in Lilac and Forsythia, or a drupe or berry in the Privet, Fringe-tree and the Olive of commerce. The last, *Olea europaèa*, L. probably native in the Mediterr. region or w. Asia, has been cult. from the earliest times in the warm-temperate regions of the old world but is not hardy in our area. Now cult. in s. Calif., Ariz., N.Mex. and Fla.

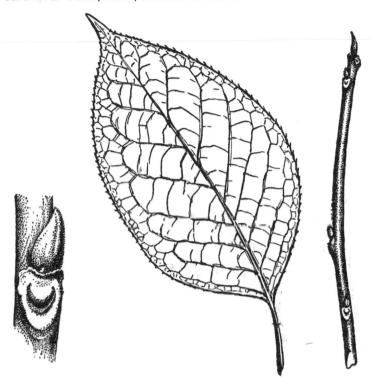

FIG. 106. *Pterostyrax hispida*, Epaulette-tree. Lf. nat. size; left, lat. bud and lf. scar × 5; right, winter brt. nat. size.

KEY TO GENERA

(Lvs. opposite for all spp.)

Lvs. in our spp. simple.
  Brts. hollow, or pith sometimes chambered (Pl. XL, 1)..............*Forsythia* 221
  Brts. with solid pith.
    Fr. a drupe.
      Small tree with white, long-petalled fls.; brts. resembling those of Ash
      (Pl. XL, 4)..*Chionanthus* 223
      Shrub with small white fls.; much used for hedges (Pl. XL, 3 and Fig. 108)
                                           Ligustrum 221
    Fr. a capsule. Commonly cult. shrubs with fragrant fls. (Pl. XL, 2)..*Syringa* 221
    Fr. a samara....................................*Fraxinus* in *F. anomala* 220
Lvs. compound; pinnate or with 3 lfts.
  Brts. hollow; pith sometimes chambered; lvs. often with 3 lfts. (Pl. XL, 1b)
                                           Forsythia 221
  Brts. with solid pith; lvs. pinnately compound with more than 3 lfts. (Pl.
    XXXIX) ....*Fraxinus* 219

PLATE XXXIX. Ashes.

1. *Fraxinus americana*, White Ash. Brt. and fr. nat. size; lf. × ⅓; terminal bud complex, side and end views × 3.
2. *F. pennsylvanica*, Red A. Brt. and fr. nat. size; lf. × ⅓ (7 lfts. often occur); lat. bud and lf. scar × 3.
3. *F. nigra*, Black A. Brt. and fr. nat. size; lf. × ⅓; lat. bud and lf. scar × 3.
4. *F. excelsior*, European A. Brt. and fr. nat. size; lf. × ½; terminal bud complex × 3.

FRÁXINUS — ASH

Lvs. pinnately compound (except in *F. anomala*, pp. 220, 221); true terminal bud present; fr. a samara or "key"; lf. scars crescent-shaped to almost circular; bundle scars numerous in a crescent- or c-shaped aggregate. The ash is sometimes confused with the hickory, but can be readily distinguished by its opposite lvs. and buds (alternate in hickory).

This key does not include *F. tomentosa*, the Pumpkin Ash.

Lvs. simple, (sometimes with 2–3 lfts.)..........................6. *F. anomala* 220
Lvs. pinnately compound; lfts. at least 5, usually more.
  Brts. pubescent (Pl. XXXIX, 2).......................2. *F. pennsylvanica* 219
  Brts. glabrous.
    Lfts. sessile, or nearly so; buds black.
      Lf. scars circular or nearly so; native tree of wet soil (Pl. XXXIX, 3)
                                 3. *F. nigra* 219
      Lf. scars semicircular; cult. only (Pl. XXXIX, 4)...........4. *F. excelsior* 220
    Lfts. stalked; buds not black.
      Buds gray; fls. attractive, white, blooming with or after lvs.; cult. (Fig. 107)
                                 5. *F. ornus* 220
      Buds rusty to dark brown, sometimes nearly black; fls. not showy, appearing
      before the lvs.; native (Pl. XXXIX, 1)................1. *F. americana* 219

1. **F. americàna** L. White Ash (Pl. XXXIX, 1). Buds stout, rusty to dark brown or sometimes nearly black; usually a pair of lateral buds very close to the base of the terminal one; *lf. scars crescent-shaped;* bark of trunk close, but grooved as in mockernut hickory; lfts. stalked, entire, undulate or serrate, may or may not be pubescent beneath; brts. smooth and shining, usually gray or greenish brown. The common sp. N.S. to Fla. and w.

The Pumpkin Ash, **F. tomentòsa** Michx. f. (*F. profunda* Bush) resembling **F. americana,** but with velvety brts., petioles and lf. rachises; and *large* fr. (2–3 in. long) has been reported from w. N.Y. State.

2. **F. pennsylvánica** Marsh. Red A. (Pl. XXXIX, 2). Buds smaller than in White A., rusty brown; *lf. scars semicircular,* not or only a little concave on upper margin; lfts. entire or wavy, stalked; *lvs. and ends of brts. downy;* smaller tree, with more slender brts, and more irregular in habit, but with bark like that of the last. Moist soil, along streams. Que. to Ala. and w.

The Green A., **Fraxinus pennsylvanica** var. **subintegérrima** (Vahl) Fern., is a var. of the Red A. in which the brts. and lvs. are glabrous (sometimes the lvs. are sparsely hairy on the veins beneath). Banks of streams. Que. to Ga. and w.

3. **F. nìgra** Marsh. Black A. (Pl. XXXIX, 3). Buds usually black and rather sharp-pointed, the first pair of lateral buds at a little distance

below the terminal bud, giving it a stalked appearance; *brts.* stout, *yellow* or buff-colored, not shining but smooth; lf. scars circular to semicircular; large trees with corky bark which easily rubs off; lfts. *sessile.* Swamps and wet soil. Nfd. to Va. and w.

4. **F. excélsior** L. European A. (Pl. XXXIX, 4). Has jet-black buds, with a pair just below the terminal bud, as in White A.; lf. scars semicircular; lfts. almost sessile, serrate. Commonly planted in parks.

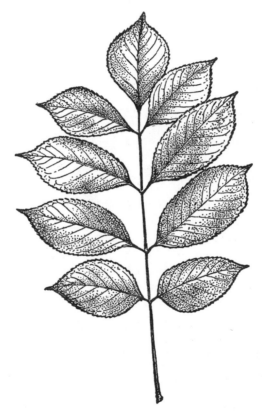

FIG. 107. *Fraxinus ornus*, Flowering Ash. Lf. × ½.

The Flowering A. (5), *Fraxinus órnus* L. (Fig. 107) has gray buds; lf. scars semi-circular or wide crescent-shaped; lfts. usually 7, oblong and abruptly narrowed into a point (the terminal one obovate) stalked, serrate, and smooth except a fringe of fine hairs along midrib beneath. In this sp. the fls. are conspicuous, white, fragrant, in dense terminal clusters; petals linear, ¼ in. long. Blooms in May and June. Occasionally cult. Eurasia.

The Single-Leaf A. (6) *F. anómala* S. Wats. is a shrub or small tree with 4-sided or even slightly ridged glabrous brts.; lvs. simple or sometimes with 2–3 lfts. Easily recognized by the characteristic *Fraxinus* buds and the simple or nearly simple lvs. Occasionally cult.   Colo. to Utah and s. Calif.

Forsỳthia

("Y" pronounced as in Thyme)

Popular, ornamental, hardy shrubs with masses of yellow fls. in early Apr.   Buds more or less cylindrical, pointed, with about 6 pairs of opposite, yellowish or yellow-brown, mucronate scales; fl. and lf. buds similar except that former are much larger and often collateral or even multiple; brts. somewhat 4-sided.   6 or 7 spp., mostly natives of China, but *F. intermèdia* is a hybrid between *F. suspénsa* and *F. viridíssima*.   The most showy form is var. *spectàbilis* of *F. intermedia*.

Brts. hollow, but solid at nodes; lvs. often with 3 lfts. or 3-parted (Pl. XL, 1, b.)
F. suspensa

Brts. when cut lengthwise, showing chambered pith.
Pith solid at nodes, elsewhere more or less chambered;* lvs. sometimes 3-parted
(Pl. XL, 1, a, c, d.)   *F. intermedia*
Pith chambered throughout; lvs. simple...................... *F. viridissima*

Syrínga — Lilac

Ornamental hardy shrubs, in one or two spp. small trees; (e. g., *S. amurensis* var. *japónica* Decne. Japanese Tree Lilac).   Buds ovoid with about 4 pairs of scales.   When these are removed, in the case of a lf. bud, the young lvs. with characteristic veining and short, wide petioles, appear.   End bud usually lacking, which causes forking of brts.; pith rather small, homogeneous, white; fls. in large clusters, fragrant, borne on brts. of the previous yr.   About 28 spp. in Asia and s. e. Eu., but a large number of vars. and hybrids.   Sp. most commonly planted is ⁰S. vulgàris, Common L., recognized by its ovate or broad-ovate and acuminate, entire, smooth lvs. (Pl. XL, 2).   The White L. is a var. For identification of the numerous spp. and vars. one should consult "Lilacs in My Garden" by Mrs. Edward Harding, or some other special work.   Living collections of spp. and vars. at the Brooklyn and N.Y. Bot. Gardens.

"*Syringa*," from the Greek *syrinx* meaning pipe, was originally applied to *Philadelphus*, the Mock-orange (p. 127) on account of its pipe-like stems, but was transferred to the Lilac genus by l'Obel, an early herbalist.   But, unfortunately, the name, as an appellation of the Mock-orange, stuck, resulting in the present confusion.   Be it known, however, that *Syringa* is now the Lilac and *Philadelphus* the Mock-orange.

Ligústrum — Privet

Deciduous, or, in our spp., often half-evergreen shrubs, holding their lvs. well into winter; buds appearing somewhat as in *Forsythia*, but much smaller, with 2 or 3 pairs of scales showing; lf. scars small, raised; brts. slender, rounded or 4-ridged below nodes; fls. small, white, in terminal clusters; but because these shrubs are much used for hedges and regularly clipped, fls. do not usually develop; fr. a drupe, black or blue-black when ripe.   Commonly planted spp. are: ⁰L. vulgàre L., Eu. Privet, with shining black fr. and young brts. minutely puberulous; hardy; ⁰L. ovalifòlium Hassk. Calif. P. (but a native

---

* Since there is great variation, we recommend cutting several shoots from different parts of the plant, before making any decision.

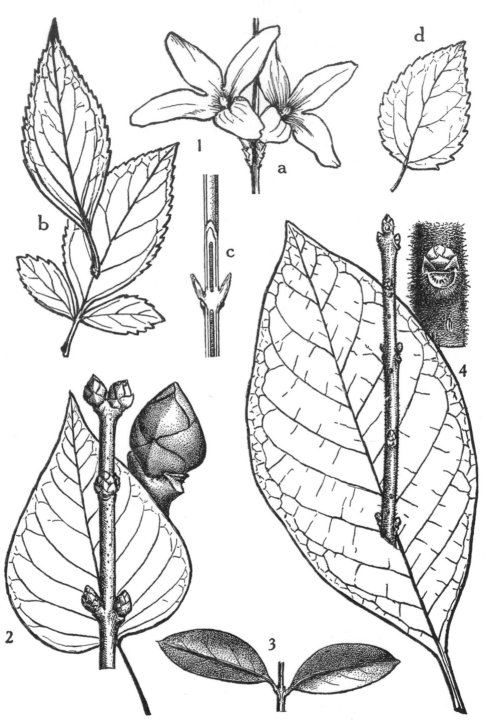

PLATE XL. Olive Family (except Ash).

1. *Forsythia.* a, fls. of *F. intermedia* nat. size; b, lvs. of *F. suspensa* nat. size; c, longisection of stem of *F. intermedia* showing solid pith at node, nat. size; d, another form of lf. of *F. intermedia* × ½. 2. *Syringa vulgaris*, Lilac. Brt. and lf. nat. size; bud × 4. 3. *Ligustrum ovalifolium*, California Privet. Part of brt. nat. size. (Lvs. below average in length). 4. *Chionanthus virginicus*, Fringe-tree. Brt. and lf. nat. size; part of brt. showing lat. bud and lf. scar × 5.

of Japan) (Pl. XL, 3 and Fig. 108), is more popular, with handsome lustrous lvs. a little wider than those of the preceding sp., and black, dull fr. — a rapid grower, but sometimes dies back in extreme cold. A var. of *L. vulgare* has bright green fr.

Fig. 108. *Ligustrum ovalifolium,* California Privet. Brt. with lvs. and fr. nat. size.

CHIONÁNTHUS — FRINGE-TREE

C. virginicus L. Fringe-tree. Old-Man's-Beard (Pl. XL, 4 and Fig. 109). Large shrub or small tree; brts. resembling somewhat those of *Fraxinus*, but lvs. simple; fls. white with narrow petals about 1 in. long, in loose, open clusters in May or June; fr. an ovoid drupe, about ½ in. long, dark blue when ripe; lvs. bright yellow in fall. N.J. to Fla. and w.

Fig. 109. *Chionanthus virginicus,* Fringe-tree, Old-Man's-Beard. Fl. brt. × ½.

LOGANIÁCEAE — LOGANIA FAMILY

BÚDDLEIA

The Butterfly-bush, ⁰*B. davidi* Franch., is a popular shrub, flowering from July until frost at a time when fls. are needed in the garden. Has opposite lvs. and long

pointed spikes of fls., each tiny fl. with an orange center. (Orangeeye B. SPN). Apt
to die back in winter, but shoots up again cheerfully and bears fls. on growth of yr.
China. Several vars. Other spp. also cult.

APOCYNÀCEAE — DOGBANE FAMILY

VINCA — PERIWINKLE

Evergreen, trailing sub-shrub, with opposite, rather leathery, shining lvs., the
horizontal stems sending up a dense mass of erect shoots to 6 or 8 in., making a good
ground cover; juice milky, a general characteristic of the family. °*Vinca mìnor* L., the
Common Periwinkle, is often used successfully as a ground cover in comparative shade.
The rather large (1 in.) blue fls. unfold from Mar. to June, often later. Cult. since
ancient times, and sometimes escaped and natzd. Eurasia. Also known as Myrtle, or
Running-myrtle, but the true Myrtle is the genus *Myrtus*, not hardy here.

VERBENACEAE — VERVAIN FAMILY

Lvs. opposite, without stipules. Trees, shrubs, or herbs. Fr. a drupe or berry.

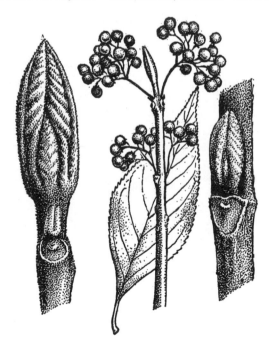

FIG. 109a. *Callicarpa japonica*, Beautyberry. Left to right, terminal bud × 4; lf. and
fr. brt. nat. size; lat. bud and lf. scar × 4.

CALLICARPA—BEAUTYBERRY

The beautyberry, °*C. japonica* Thunb., a shrub to 8 ft., with elliptic to ovate-
lanceolate, acuminate, serrulate lvs., glandular beneath, and buds often stalked, naked,

or the smaller ones with 2 nearly valvate scales and with stellate pubescence, is occasionally cult. and is becoming popular. Best recognized in the winter by its plentiful clusters of violet-colored berry-like drupes, each about the size of a tapioca grain. Fr. persistent through winter and attractive. C. China (Fig. 109a).

VITEX — CHASTE-TREE

The Chaste-tree, ⁰V. agnus-castus L., a shrub to 9 ft., native of s. Eu. and w. Asia, with opposite, palmately compound lvs., the 5–7 lfts. short stalked, lanceolate and long-pointed, small often superposed buds, u-shaped lf. scars, 4-sided brts. and purple fls. in long spikes in late summer and early autumn, is sometimes escaped in s. N.E. Whole plant pubescent with strong, aromatic odor. Matrons of ancient Greece are said to have strewn their beds with this plant. S.e. Eu. and w. Asia.

LABIÀTAE — MINT FAMILY

The Germander, Teùcrium; Rosemary, Rosmarìnus; Lavender, Lavándula; Sage, Sálvia; Thyme, Thỳmus; and others, are all more or less woody or contain woody spp., but of comparatively small size. They are often cult., especially in "herb gardens." As members of this large family (most of which are herbaceous) they have opposite lvs., with an aromatic odor, stems more or less square (in cross-section) and "two-lipped" fls. (see under Lonicera, Honeysuckle, pl. XLV, 2), hence the name "Labiatae"; and the fr. consists of 4 nutlets or achenes.

FIG. 110. Lycium halimifolium, Matrimony Vine. Fl. brt. nat. size.

SOLANÀCEAE — NIGHTSHADE FAMILY

SOLÀNUM — NIGHTSHADE

⁰⁰S. dulcamàra L. Bitter Nightshade or Bittersweet (Pl. XLI, 1). A woody climber, natzd. in some places, with clusters of bright red berries,

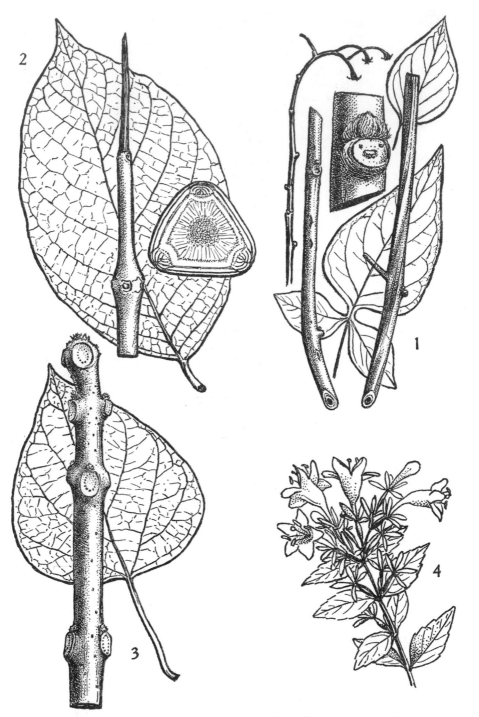

PLATE XLI. Bitter Nightshade, Buttonbush, Catalpa, and Abelia.

1. *Solanum dulcamara*, Bitter Nightshade, Bittersweet. 2 forms of lvs. and 3 forms of brts. nat. size; part of brt. showing lf. scar and lat. bud × 5.

2. *Cephalanthus occidentalis*, Buttonbush. Brt. with dead tip and lf. nat. size; cross section of stem at node showing 3 buds imbedded in bark × 5.

3. *Catalpa bignonioides*, Common C. Brt. nat. size; lf. × ⅓.

4. *Abelia grandiflora*, Glossy Abelia. Flowering brt. nat. size.

poisonous when eaten.   Lvs. simple, alternate, ovate or heart-shaped, or with two ear-like lobes at the base which sometimes become separate lfts.; stems light gray or greenish, soon hollow, with greenish wood, usually somewhat downy, striate, or irregularly 3-sided, rarely prickly; buds globose, alternate; lf. scars raised, semicircular.   Not a relative of the Climbing Bittersweet.   Natzd. from Eurasia.

Lýcium

⁰**L. halimifòlium** Mill.   Matrimony Vine (Fig. 110).   Shrub, with long recurved more or less thorny, angled, light gray brts., lanceolate or oblong-lanceolate lvs.; greenish-purple, short, funnel-form fls.; and orange-red or scarlet small berries, often in 2's — this last said to be the reason for the common name.   Cult. and sometimes escaped.   Eu.

SCROPHULARIÀCEAE — FIGWORT FAMILY

Paulòwnia

P. tomentòsa (Thunb.) Steud.   Paulownia, Princess-tree (Pl. XLII). Lvs. opposite, heart-shaped, soft pubescent; often showing very shallow lobing, indicated by projecting points on the margins; large, handsome, violet, unequally 5-lobed fls. in upright panicles in May before the lvs.; clusters of large, ovoid capsules evident in winter, as well as clusters of fl. buds for the following yr.; seeds small, winged; lf. buds blunt, sunken in bark, superposed above large, nearly circular lf. scars; terminal bud lacking; pith usually chambered, or hollow, white.   China.   Escaped and natzd. from N.Y. to Fla. and Tex.   Not hardy n.

BIGNONIÀCEAE — BIGNONIA FAMILY

Campsis — Trumpet-Creeper

⁰⁰**C. radìcans** (L.) Seem.   Trumpet-creeper, Trumpet-vine (Fig. 111). A high-climbing shrub with opposite, pinnately compound deciduous lvs. and conspicuous orange-scarlet tubular fls., climbing mainly by aerial rootlets, produced in double bands below the nodes; fr. a long (3 in. or more) thick pod, somewhat like that of Catalpa but shorter and much thicker; lf. scars elliptical or shield-shaped; buds small with 1–3 pairs of scales showing; lf. scars more or less connected by hairy transverse ridges.   N.J. and e. Pa. to Fla. and w. but natzd. n. to Conn.   Handsome in cult. In southern states an aggressive pest.

Catálpa

Apt to be confused with Paulownia, but has *lvs. usually 3 at a node* (sometimes 2), smoother, usually without a tendency toward lobing, and

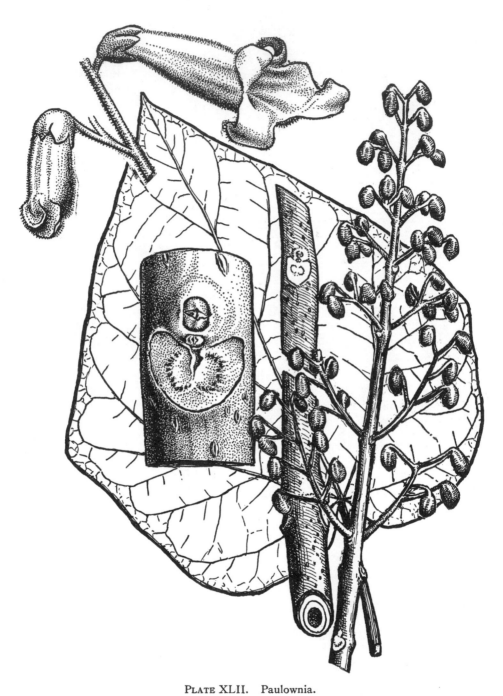

PLATE XLII.   Paulownia.

Lf., fl. buds, fls. and brt. nat. size; part of brt. with lat. bud and lf. scar × 3.

with *solitary buds* in their axils; nearly white, spotted fls., opening later than in Paulownia; very long, nearly *cylindrical* usually *persistent pods*, with large winged seeds.

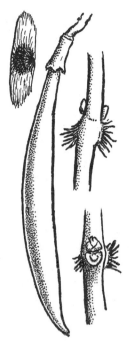

Fig. 111. *Campsis radicans*, Trumpet-creeper. Brts. with lf. scar, buds and adventitious roots (used in climbing), fr. pod, and seed, all nat. size.

C. speciòsa Warder. Northern Catalpa, Hardy C. Fls. (in June) *inconspicuously spotted;* lvs. long-pointed, pubescent beneath, without odor when crushed; fr. 8–20 in. long, ½–¾ in. *in diam. at the middle; wings of seeds rounded* at end, with fringe of short hairs. Native in the Middle Western U. S. Cult. and escaped and natzd. in Eastern States.

C. bignonioìdes Walt. Southern Catalpa, Common C. (Pl. XLI, 3). Fls. later (in June and July) *conspicuously spotted;* lvs. pubescent beneath, short-pointed, these and brts. with unpleasant odor when crushed; fr. 6–16 in. long, ⅓–½ in. *in diam. at the middle; wings of seeds pointed* at end, with a fringe of long hairs. S. U.S. Cult. and often escaped in n.

Differences between the 2 Common Spp. of Catalpa

	C. speciosa	C. bignonioides
1. Habit	Pyramidal	With spreading branches
2. Height	To 90 ft.	To 45, rarely 60 ft.
3. Lf. shape	With long, tapering points	Abruptly short-pointed

4. Lf. and brt. odor	No unpleasant odor when crushed	With unpleasant odor when crushed
5. Fls	Inconspicuously spotted Corolla larger, 1½–2 in. long	Conspicuously spotted Corolla smaller, 1–1½ in. long
6. Fr	½–¾ in. thick at middle Valves remaining rounded after separating	⅛–½ in. thick at middle Valves flattening after separating
7. Seeds	Wings and hairs rounded at ends	Wings and hairs coming to a point

C. ovàta Don., Chinese C., with *glabrous* lvs., often lobed, and yellow fls., orange striped and violet spotted, is sometimes cult.; fr. pods ⅕–⅓ in. thick.

RUBIÀCEAE — MADDER FAMILY

CEPHALÁNTHUS — BUTTONBUSH

C. occidentàlis L.   Buttonbush (Pl. XLI, 2).   Usually a shrub, with ovate, entire lvs., opposite or in 3's or 4's, with triangular stipules; small buds in depressed areas and surrounded by outer bark, above the circular lf. scars; brts. rounded, dying back or bearing fls.; pith light brown, more or less 4- or 6-sided; fls. small, white, in July and Aug. in dense, spherical heads, and small, dry, persistent frs. clustered the same way; wet soil or swamps.   N.S. to Fla. and w.

MITCHÉLLA — PARTRIDGEBERRY

The Partridgeberry, ⁰**M. rèpens,** is a small, creeping plant on the border line between shrub and herb, easily recognized by the scarlet frs. in united pairs and by the small, rounded, opposite, evergreen lvs.; frs. persistent and attractive all winter; edible but tasteless.   Much used for rockeries and dish gardens.   Woods.   Nfd. to Fla. and w.

CAPRIFOLIÀCEAE — HONEYSUCKLE FAMILY

Entire family has opposite lvs. (and buds)

KEY TO GENERA

SAMBÙCUS — ELDER

Lvs. pinnate; pith wide and soft; true terminal bud lacking; stipule
scars lacking; lf. scars meeting laterally or connected by transverse lines.

⁰S. canadénsis L. American Elder (Pl. XLIV, 1). Pith white; fls.
and frs. in flat-topped clusters. The common sp., with black berries.
N.S. to Ga. and w.

⁰S. pùbens Michx. Red-berried E. (Pl. XLIV, 2). Pith *brown or
orange;* buds and young brts. often with purplish tinge; fls. and frs. in
elongated clusters; berries red, and very striking. Rocky woods. Nfd.
to Ga. and w.

VIBÚRNUM

Likely to be confused with *Cornus* because of the opposite, simple lvs.
but they are nearly always *toothed* or *lobed*, while in the Dogwoods they
are entire. (In **V. cassinoides** and **nudum** they are sometimes entire).
As in *Cornus*, the fls. are small and white, in clusters which may be flat,
pyramidal, or rounded. In some spp., e. g. **V. alnifolium**, the marginal
fls. are enlarged and sterile, and in two vars. *all* fls. are of this kind and
the whole cluster is rounded — i.e. the Japanese Snowball, *V. tomento-
sum* var. *sterile* and the Snowball-tree, a var. of the Guelder-Rose, i. e.,
V. opulus var. *roseum*. Ripe fr. is a black or very dark blue drupe (red
in *V. opulus*, **trilobum** and some oriental spp.); true terminal bud present; stipule scars lacking; lf. scars nearly or quite meeting at sides
(sometimes joined by a transverse ridge): buds naked or covered by 2
valvate or 2 or 3 pairs of scales. For an excellent article on the ornamental value of Viburnums see P. & G. 7: 12–15. 1951.

Lvs. 3-lobed.
  Fl. clusters without enlarged sterile marginal fls.; fr. blue-black (Fig. 112)
  Fl. clusters with enlarged sterile marginal fls. (or in *V. opulus* var. *roseum*, all
      fls. sterile in round head); fr. red.
    Petioles with small, dome-shaped or columnar glands; lvs. glabrous, except
Lvs. not lobed.
  Fl. cluster pyramidal or rounded; bud scales often connate; pubescence stellate;
  Fl. cluster umbel-like, more or less flat-topped (except in "Snowball" form of *V.*
      tomentosum).
    Buds naked, i. e. without true bud scales; pubescence stellate.
      Marginal sterile fls. present and conspicuous, to 1 in. wide; native in rich
      No marginal sterile fls.; cult.
        Lvs. broad-ovate to elliptic, 1–4 in. long, irregularly toothed; petioles ¼–
        Lvs. ovate to oblong-ovate, with closely denticulate margins; petioles ½–1
    Buds scaly.
      Fl. clusters with large sterile marginal fls. (in "Snowball" all fls. enlarged
      Fl. clusters without sterile fls.; pubescence (when present) not stellate.
        Lf. veins curving and joining before reaching margin of lf.
          Fl. cluster raised on a central basal stalk (Pl. XLIII, 1)
          Fl. cluster without a stalk (sessile).
            Petiole with broad, wavy margin, brts. slender, buds very long-
            Petiole without wavy margin, brts. stiff, at wide angles and usually
        Lf. veins straight, ending in the teeth.
          Brts. glabrous; lvs. glabrous beneath except in axils of veins (Pl. XLIII,
          Brts. slightly pubescent; lvs. pubescent beneath, sometimes only

1. ⁰**V. acerifòlium** L. Maple-leaved Viburnum (Fig. 112). Low
shrub; lvs. pubescent, toothed, shaped like those of red maple, with
minute dark dots on under surface; bud scales separate, i. e. not entirely
valvate, the outermost pair very short. Common in woods. Magnificent fall purple coloring. Que. to Ga. and w.

2. ⁰**V. trilòbum** Marsh. Highbush-Cranberry. Amer. Cranberry-
bush. Large shrub or small tree to 12 ft. with gray bark; buds with 2
outer scales connate; lvs. 3-lobed, glabrous; petioles glandular where they
join lf.; the glands *convex, club-shaped* or *columnar;* fr. clusters orange to

red with taste like that of cranberries.   Much used for preserves and jelly.   Nfd. to Pa. and w.

The Eu. Cranberry-bush 3, ⁰ *V. ópulus* L., Guelder-Rose, is very similar, but lf. lobes are more rounded; petiolar glands are *concave-tipped*; fr. clusters smaller and fr. not so pleasant, sometimes bitter.   Sometimes escaped.   The Snowball-tree, var. *ròseum*, has all fls. enlarged and sterile.

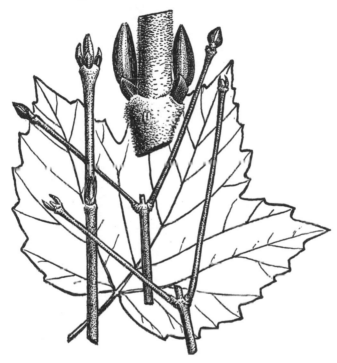

Fig. 112.   *Viburnum acerifolium*, Maple-leaved Viburnum.   Lf. and brts. nat. size; pair of lat. buds on brt. × 4.

4. V. siebòldi Miq.   Siebold V. (Fig. 113).   Shrub or small tree to 30 ft.; brts. stout; bud scales valvate, often connate, buds more or less 4-sided; fl. buds larger, ovoid; lf. scars triangular, nearly or quite meeting; buds, brts., and lf. underside, (especially the veins) with stellate pubescence; lvs. large, 2–6 in., oblong obovate and more or less rounded, shining above with veins *much impressed;* frs. ripening through pink and red to blue-black; fl. clusters showy white, in May or June.   Young lvs. when crushed have odor (to author at least) of burned bacon.   Japan. One of most attractive spp.

5. ⁰V. alnifòlium Marsh.   Hobblebush (Fig. 114).   Buds large and naked, scurfy; lvs. large, broad ovate or nearly circular (4–8 in.); stellate

FIG. 113.   *Viburnum sieboldi*, Siebold Viburnum.   Lf. and part of brt. nat. size.

pubescent or scurfy beneath, especially on veins; fl. and fr. clusters 3–6 in. broad, outer fls. sterile and showy, to 1 in. broad.   Woods and cool ravines, often with *Acer spicatum* and *A. pensylvanicum*.   P.E.I. to Ont. s. to n. N.J. and Pa., upland to Ga. and w.

6. ⁰V. cárlesi Hemsl.   Mayflower V., Fragrant V.   Low shrub to 5 ft.; buds naked; fl. buds often with a pair of opposite scale lvs. projecting at a wide angle like horns, at tip of brts., scurfy; fls. pink, salver-shaped, *very fragrant;* lvs. broad ovate to elliptic, 1–4 in. long, stellate pubescent above and below.   Increasingly popular because of its fragrant and handsome early fls.   Korea.

The Wayfaring-tree 7, *Viburnum lantana* L., is an ornamental shrub with lvs. resembling somewhat those of **V. alnifolium** (fig. 114), but smaller; fls. in umbel-like clusters without enlarged sterile ones, and stellate-pubescent, scurfy brts.   Eurasia.

8. ⁰V. tomentòsum Thunb.   Double-file V.   Shrub to 10 ft.; brts. horizontally spreading, stellate tomentose; lvs. broad ovate to oblong ovate—to 4 in. long, stellate pubescent beneath; buds with valvate often connate scales; fr. red, ripening to blue-black; fl. clusters flat with marginal fls. sterile, but in var. *sterile*, Japanese Snowball, rounded, all the fls. being showy and sterile.   Japan, China.

9. ⁰V. cassinoìdes L.   Withe-rod, Wild-raisin (Pl. XLIII, 1).   Shrub to 6 ft.; buds yellow-brown or golden; lvs. dull above, crenulate, dentate or entire with short, blunt point.   Resembles **V. prunifolium** and **len-**

tago, but whole fl. cluster is *stalked;* fr. sweet.   More common n. and upland.   Nfd. to Ala. and w.

The Possum- or Swamp-Haw 10, **V. nùdum** L., like the Withe-rod, has a definitely stalked fl. and fr. cluster, but is a larger, coarser plant, sometimes tree-like, to 18 ft. with a trunk 4–8 in. in diam.: lvs. shining above and nearly entire; brts. shining and fr. often bitter; buds red-brown or tawny.   A more southern sp. than **V. cassinoides.** Wooded swamps and bogs; s. Conn. to Fla. and w.

11. **V. lentàgo** L.   Nannyberry (Pl. XLIII, 3).   Shrub or small tree to 30 ft.; lvs. acuminate, *with winged petioles;* bud scales nearly or quite valvate; fl. buds swollen at base; lf. buds long and narrow; drupes ellipsoid, blue-black and sweet; heart wood has nasty odor; called, locally, "stink-wood."   Que. to Pa., upland to Ga. and w.

12. **V. prunifòlium** L.   Black-haw (Pl. XLIII, 4).   Shrub or small tree to 20 ft.; with short, stiff brts. often nearly at rt. angles to stem; bark somewhat like that of *Cornus florida;* lvs. oval, finely serrate; bud scales valvate or nearly so.   Conn. and N.Y. to Fla. and w.

13. ⁰**V. dentàtum** L.   Arrow-wood (Pl. XLIII, 2).   Lvs. with coarse teeth, nearly smooth beneath; bud scales separate, ciliate, the outermost pair rather short, but sometimes reaching the middle of the bud; lf. scars ciliate.   Common in wet or moist ground.   (*V. dentatum* of ed. 7 of Gray's Manual. *V. recognitum* Fern.).   N.B to S.C. and w.

FIG. 114.   *Viburnum alnifolium,* Hobblebush, Witch-hobble.   Lf., below average size.

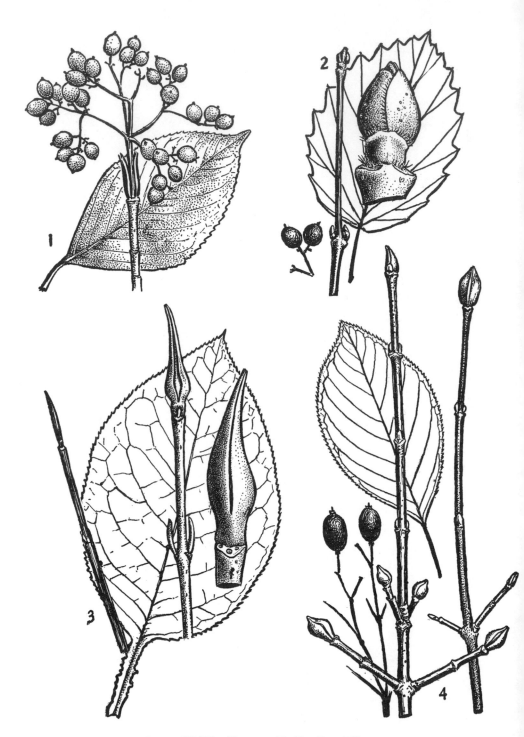

PLATE XLIII.   Honeysuckle Family.   Viburnums.

1. *Viburnum cassinoides*, Withe-rod.   Lf. and fruiting brt. nat. size.   2. *V. dentatum*, Arrow-wood.   Lf., brt. and fr. nat. size; terminal bud × 5.   3. *V. lentago*, Nannyberry.   Lf., brt. with fl. and lf. buds, and vegetative brt. with lf. buds only, nat. size; fl. bud × 2.   4. *V. prunifolium*, Black-haw.   Lf., brts. with lf. and fl. buds; frs. and remains of fr. cluster; all nat. size.

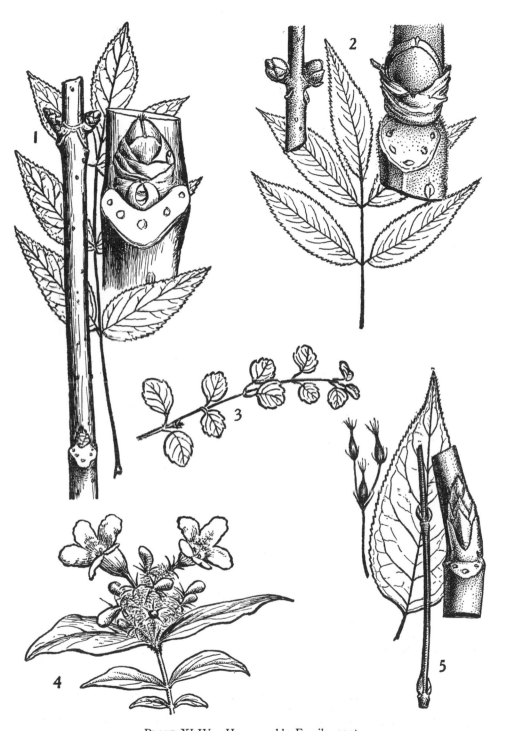

PLATE XLIV. Honeysuckle Family cont.

Elders, Twinflower, Beautybush and Bush-honeysuckle.

1. *Sambucus canadensis*, Common Elder.   Brt. nat. size; lf. × ½; part of brt. with lat. bud and lf. scar × 4.   2. *S. pubens*, Red-berried E.   Part of brt. nat. size; lf. × ½; part of brt. with lat. bud and lf. scar × 4.   3. *Linnaea borealis*, var. *americana*, Twinflower.   Brt. nat. size.   4. *Kolkwitzia amabilis*, Beautybush.   Flowering brt. nat. size.   5. *Diervilla lonicera*, Bush-honey-suckle.   Lf., brt. and cluster of frs. nat. size; part of brt. with lat. bud and lf. scar × 5.

The Downy Arrow-wood 14, ⁰**V. pubéscens** Pursh. of ed. 7 of Gray's Manual, which closely resembles **V. dentatum,** but has lvs. pubescent beneath and grows on higher, drier ground, is included by Fernald in the 8th ed. of Gray's Manual as *V. rafinesquiànum* Schultes. Dry slopes, open woods etc. Que. to Ga. and w.

The Burkwood Viburnum, *V. burkwoodi,* a hybrid (1924) of *V. utile* and *V. carlesi,* is regarded as an improvement over the deliciously fragrant and desirable *V. carlesi* because more vigorous and easily propagated by cuttings. *V. carlesi* is said to be subject to a disease incident to grafting (61).

SYMPHORICÁRPOS

Small shrubs; buds small with 2 or 3 pairs of keeled bud scales; lf. scars more or less connected by transverse ridge.

1. ⁰**S. álbus** (L.) Blake. Snowberry. Brs. very slender, old bark shreddy; brts. hollow; lvs. oval, small, 1–2 in. long; fls. pinkish; *berries white*, globose, to ½ in. in diam. and conspicuous in early winter. Que. to Va. and w. The form usually cult. is *S. albus laevigatus* (Fern.) Blake.

2. ⁰**S. orbiculàtus** Moench. Coralberry, Indian-currant. Similar; pith solid; berries purplish-red in dense, elongated clusters, somewhat smaller than in last. Pa. and s.; cult. and sometimes escaped.

ABÈLIA

⁰*A. grandiflòra* (André) Rehd. (Pl. XLI, 4). A hybrid of 2 spp.—*A. chinensis* and *A. uniflora*—with small, dark green lvs. and flowering from June to Nov. Cult. for its long flowering season, and its lustrous, half-evergreen lvs. which become bronzed in fall.

LINNAÈA — TWINFLOWER

⁰**L. boreàlis** (L.) var. **americàna** (Forbes) Rehd. (Pl. XLIV, 3). Small, exquisite, trailing plant, evergreen, useful in rockeries in shady positions. Cool woods. Greenl. to W. Va. and w.

KOLKWÍTZIA

⁰*K. amàbilis* Graebn. Beautybush (Pl. XLIV, 4). Upright shrub; buds with several pairs of pointed, pubescent scales; lf. scars crescent-shaped or triangular, connected by transverse lines; young brts. bristly-hairy; fls. pink, borne profusely in June; fl. clusters and fr. bristly-hairy. Cult. China.

DIERVÍLLA — BUSH-HONEYSUCKLE

⁰**D. lonícera** Mill. Dwarf Bush-Honeysuckle. (Pl. XLIV, 5). Lvs. oblong-ovate, taper-pointed, serrate; yellow to red fls. ripening into slender, long-pointed, persistent, dry frs.; pointed, scaly buds; terminal bud present; stipule scars lacking. Nfd. to N.C.

WEÌGELA

⁰*Weigela* SPN. (often called *Weigèlia*) is similar to the last and is a commonly cult. shrub in the sp. *W. flòrida* (Sieb. and Zucc.) A. DC. (Pl. XLV, 1, a, b). Has attractive funnel-shaped fls. about 1 in. long. Several vars. with rosy-pink or red fls.; frs. beaked at first, at length blunt. N. China and Korea.

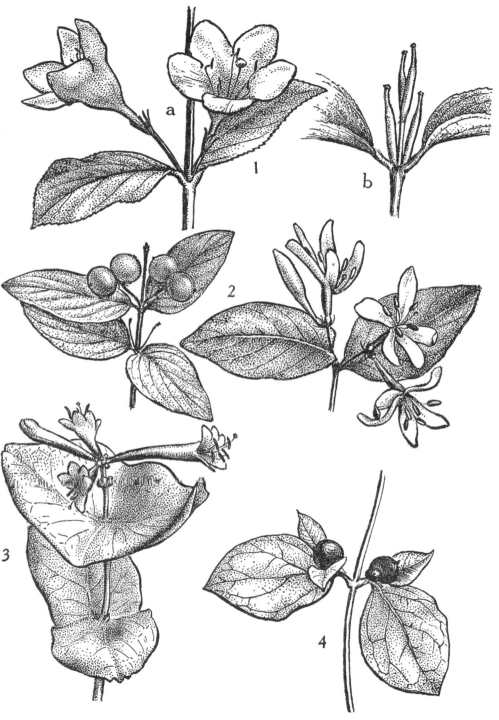

PLATE XLV. Honeysuckle Family concl.
Weigela and Honeysuckle.

1. *Weigela florida*, Weigela. a, Fl. brt.; b, fruiting brt. nat. size.
2. *Lonicera tatarica*, Tatarian Honeysuckle.   Rt., fl. brt.; lft., fr. brt.; nat. size.
3. *L. sempervirens*, Trumpet H.   Fl. brt. nat. size.
4. *L. japonica*, Japanese H.   Fr. brt. nat. size.

Lonícera — Honeysuckle

Shrubs or woody vines; lvs. opposite, simple, entire, sometimes connate; fls. (and fr.) in axillary pairs or in whorls at or near ends of brts. A large genus, but easily recognizable from its simple, opposite lvs., and its more or less 2-lipped fls., i.e. the 5 lobes of the fl. in 2 groups—3 and 2 (Pl. XLV, 2). The following key may serve as a guide to the commoner spp.—Rehder's Manual should be consulted for further description and for other spp.—also Gray's Manual 8th ed. for the native or natzd. spp. The Japanese H. is so well established here that in many places it is a pest, especially on Long Island, N.Y. The Tatarian H. with its many vars. is most common in cult. and an old reliable. A very fine kind, especially for its red autumn fr., is *L. maacki* var. *podocárpa*. *L. fragrantissima*, the Winter H. with its early fls. (sometimes even in Jan., in Brooklyn, N.Y.) is valuable for its fragrance—a delicious, fruity odor—and its earliness.

Fls. in pairs, axillary.
  Brts. with solid pith.
    Fls. with nearly regular lobes, nodding on slender stalks; yellowish; lvs. ovate, ciliate; native small shrub (to 5 ft.) in woods........ᵒ**L. canadénsis** Bartr.
                                                       Fly Honeysuckle
    Fls. 2-lipped; white or yellowish, fragrant, early blooming (Jan.–Apr.) lvs. elliptic to broad-ovate, ciliate (1–3 in. long); cult. shrub..ᵒ*L. fragrantissima* Lindl. and Paxt.
                                                        Winter H.
  Brts. hollow.
    Upright shrub; fls. light pink (deeper pink or red in some vars.); fr. red (Pl. XLV, 2)..ᵒ*L. tatárica* L.
                                                      Tatarian H.
    Twining vine; fls. white, changing to yellow; fr. black; sometimes climbing over other shrubs, often trailing on the ground (Pl. XLV, 4). Natzd. from Asia..ᵒᵒ**L. japónica** Thunb.
                                                      Japanese H.
Fls. in whorls at or near ends of brts.
  Lvs. all with stalks; fls. two-lipped, forming dense heads, yellowish white, often tinged with purple; twining, natzd. vine........ᵒᵒ*L. periclýmenum* L.
                                                      Woodbine H.
  Lvs. below fls. usually connate.
    Twining native vine; fls. red, trumpet shaped, with short lobes; lvs. evergreen in s. (Pl. XLV, 3)...............................**L. sempervìrens** L.
                                          Trumpet or Coral H.
    Weakly reclining native shrub; fls. deeply 2-lipped, green or yellowish
                                              L. dioìca L.
                                              Limber H.

The Morrow Honeysuckle, ⁰*Lonicera morrowi* A. Gray, a shrub to about 6 ft. with deeply 2-lipped fls. white changing to yellow, small, obtuse winter buds and dark red fr., soft pubescent young brts., is often cult. and sometimes natzd. in our area. Lvs. elliptic to ovate-oblong, acute and mucronulate.   Very hardy, Japan.

COMPÓSITAE — COMPOSITE FAMILY

The largest family of angiosperms, about 1100 genera, and more than 20,000 spp. but mostly herbs.   Fls., e.g., in the common white daisy, *composed* of tiny individual fls. in close arrangement to form a composite head, which, as a whole seems like a single fl.; some of the tiny fls. with long petal-like extensions, the "ray fls.," and others, forming the central "disk," without these showy parts.   The ray fls. serve to attract the atten-

FIG. 115.   *Baccharis halimifolia*, Groundsel-tree or -bush.   Lvs. and fr. brt. showing white pappus.   Nat. size.

tion of insects, which bring about cross pollination, while the disk fls. specialize in the formation of seeds.   In other members of the family, as in Dandelion, Thistle, etc., all fls. of the head are the same.   The **Compositae** are believed to represent the most advanced form, to date, of the main line of plant evolution.

Most of the woody spp. grow in tropical or subtropical regions, but the 2 following shrubs are of common occurrence in salt marshes or in thickets at their margin.

BÁCCHARIS — GROUNDSEL-TREE

The Groundsel-tree, ⁰**B. halimifòlia** L. (Fig. 115), to 12 ft., is distinguished by its alternate, obovate, simple lvs., bluntly toothed in the upper part (upper lvs. entire, or nearly so); slender, green or brownish, ridged brts. with minute, resinous buds, and broadly v-shaped lf. scars.   Stipule scars absent.   Since the brts. have indeterminate growth, a typical terminal lf. bud is lacking.   Fls. in terminal clusters, dioecious, the pistillate plants conspicuous in late fall with their tassels of white pappus.   Near the coast, Mass. to Fla., Tex., and Mex.

FIG. 116.   *Iva frutescens*, var. *oraria*, Marsh-elder.   Fl. brt. × ½.

ÌVA — MARSH-ELDER

The Marsh-elder, ⁰**I. frutéscens** L. var. **orària** (Bartlett) Fern. and Grisc. (Fig. 116), is lower (2–6 ft.), has longer, oval or lanceolate somewhat fleshy, *sharply toothed*, mostly *opposite lvs.*; ridged brts.; and fls. resembling those of the ragweed; lf. scars broadly triangular, and, when opposite, connected by lateral flaps.   Stipule scars absent.   Salt marshes, N.H. to Va.

SHORT CUTS TO NAMING

In case time is short and the "flesh is weak," I have given below a few simple ways to identify some of the woody plants. For there are many kinds (not all) which have some unusual feature by which they can be immediately recognized.

But first a few general tips. If your plant belongs to the Pine Family, with mostly evergreen and needle-like or scale-like lvs. and the fr. in cones, its identification should be easy: the key is short and includes also the Yew Family (pp. 13, 14).

If the lvs. are opposite[1] I recommend using the keys (pp. 14–34) for there are comparatively few plants with opposite lvs. Certainly if the lvs. are opposite and evergreen, naming is simple (p. 14), and also if the lvs. are evergreen and alternate (pp. 18, 19) there should be little difficulty. But if the lvs. are deciduous and alternate the task is much harder. (See D. Broad-lvd. woody plants with alternate deciduous lvs. pp. 20–29 + 34.)

May I emphasize the fact that the following "short cuts" do not include *all* of the plants—only some with unusual and striking characteristics. Many of the different groups are only roughly arranged in the form of keys, so that one will have to read down until he finds the description that fits his plant.

CHARACTERS OF BUDS

I. LARGE AND CONSPICUOUS TRUE TERMINAL BUDS
1. Hickory. Lvs. alternate, pinnately compound, pith solid, angled (Pl. VII and p. 85).
2. Horse-chestnut. Lvs. opposite, palmately compound (Fig. 80 and p. 176). See also other spp. of *Aesculus* (Fig. 81 and pp. 175, 176).
3. Magnolia. Lvs. alternate, simple, entire; brts. aromatic; pith chambered or diaphragmed (Pl. XVIII and fig. 42 and pp. 120–123).
4. Sweet-gum. Lvs. star-shaped, alternate; brts. with corky ridges (Pl. XIX, 3 and p. 129).
5. Tulip-tree. Lvs. alternate, lobed, squarish at tip; buds flattened like duck's bill; pith chambered or diaphragmed (Fig. 43 and pp. 123, 124).
6. Walnut and Butternut. Lvs. alternate, pinnate; pith chambered (Pl. VI and pp. 83, 85).
Maple (Pl. XXVIII) with opposite, usually palmately lobed lvs., and Ash (Pl. XXXIX) with opposite, pinnate lvs., might also fall into this category, although their buds are not conspicuously large.
II. BUDS LONG-POINTED AND SLENDER (Lvs. simple, alternate, serrate)
1. Beech. Buds brown or red-brown, ¾–1 in. long; lvs. pinnately straight-veined; bark of old trunks *smooth* (Pl. XII and pp. 96, 98).

[1] Assuming that the plant is *not* in the Pine Family.

2. Birch. Buds light brown, about ¼ in. long; lvs. doubly serrate; bark with wintergreen taste and prominent lenticels (Pl. VIII and p. 89).

3. Shadbush. Buds greenish, ½–⅝ in. long; lvs. simple, serrate; bark smooth; shrubs or small trees (Fig. 59 and pp. 139, 140).

III. BUDS WITH ONLY ONE CAP-LIKE SCALE

1. Sycamore and London Plane. Lvs. lobed, alternate; base of petiole covering bud (Fig. 49, p. 131, and p. 132).

2. Willow. Lvs. simple, alternate (except sometimes opposite in Purple-osier W.); mostly shrubs but sometimes trees (Pl. III and figs. 25, 26, 27, 28, and pp. 74–78).

IV. BUDS (AT LEAST END BUDS) NAKED, i.e., WITHOUT TYPICAL SCALY COVERINGS

1. Glossy Buckthorn. Small tree with simple lvs.; purple frs. (drupes) (p. 176).

2. Franklinia. Rare tree (p. 181).

3. Pawpaw. Terminal bud like that of Witch-Hazel, but lvs. larger (6–12 in. long) (p. 126).

4. Viburnums. Hobblebush, native, Mayflower V., cult., and Wayfaring-tree, cult. Lvs. opposite; fls. white or pinkish, in large flat-topped clusters (Fig. 114 and pp. 233, 234).

5. Witch-Hazel. Bud flat with knife-like outline; lvs. oval, wavy-toothed, one-sided at base (Pl. XI, 3 and p. 129).

V. WITH NAKED CATKINS FALL AND WINTER

1. Alders. Shrubs; catkins reddish (Pl. X, 1, 2, and pp. 91–94).

2. Birches. Mostly large trees (Pl. VIII, IX, and pp. 87–91).

3. Hazelnuts. Shrubs; catkins gray (Pl. XI, 1, 2, and p. 96).

4. Hop-Hornbeam. Medium-sized trees; bark in long, irregular strips; catkins in 2's or 3's (Pl. X, 5, and p. 94).

VI. BUDS CLUSTERED AT ENDS OF BRTS.

1. Azalea. Shrubs (Pl. XXXIII, 2, fig. 95, and pp. 198–200).

2. Oak. (Pls. XIII, XIV, and figs. 32, 33, and pp. 99–110).

VII. LEAF SCARS ALMOST OR QUITE SURROUNDING BUDS OR COVERING THEM

1. Mock-orange. Lvs. opposite, simple; shrubs with large, white-petalled, usually fragrant fls. (Figs. 45, 46, p. 127, and p. 221).

2. Leatherwood. Lvs. alternate, simple, entire (Pl. XXX, 4, Fig. 87, p. 184 and p. 185).

3. Sumac spp. Lvs. alternate, compound (Pls. XXIV, XXV, and pp. 161–165).

4. Sycamore and London Plane. Lvs. alternate, palmately lobed (Fig. 49, p. 131, and p. 132).

5. Yellowwood. Lvs. alternate, compound. Tree, with smooth bark. (Pl. XXI, 1, and p. 153).

CHARACTERS OF STEMS

I. BRTS. WITH UNUSUAL COLORING

1. *Bright Green*

(1) Euonymus spp. Lvs. opposite; fr. usually 4-lobed (Pl. XXVII, 4, Fig. 77, a, p. 169, and pp. 169, 170).

(2) Kerria. Ornamental shrub, lvs. alternate, simple, with single or double, bright yellow fls. (Fig. 63, p. 142 and p. 142).

(3) Sassafras. Brts. aromatic: lvs. alternate, simple, lobed) (Pl. XIX, 2, and p. 126).

(4) Smilax. Prickly vine (Fig. 24, p. 73, and p. 73).

  2. *Dark Green*

   (1) Japanese Pagoda-tree. Tree of Pulse Family. Lvs. pinnate, alternate; fls. pale yellow in pendent clusters (Fig. 71 and p. 153).

   (2) Striped Maple. Lvs. lobed, opposite; brts. and trunk with white stripes (Pl. XXVIII, 4 and p. 175).

  3. *Bright Yellow*

   (1) Yellow-twig Dogwood. Lvs. opposite, entire; ornamental shrub (p. 194).

   (2) Weeping Willow. Lvs. alternate.

    Two kinds are common:

     Weeping White Willow, *S. alba tristis* (p. 77, no. 3), and Babylon Willow, with more olive-colored brts. (p. 77, no. 5).

  4. *Various Shades of Green, Pink, Red, and Purple* (See *Cornus*, pp. 189–195).

II. TRUNKS WITH UNUSUAL COLORING

  1. *White Bark* (Lvs. alternate, serrate)

   (1) Gray Birch. Lvs. very long-pointed; outer bark rather dull-colored, not peeling off readily. Small tree often in clumps (Pl. VIII, 4, and pp. 89–91).

   (2) Paper Birch. Lvs. ovate; outer bark peeling off in very thin sheets, white and chalky. Large tree with usually one main trunk (Pl. IX, 3 and p. 91; see also European White B. p. 91, nos. 7 and 8).

  2. *With White Patches*

   Sycamore. Lvs. lobed; petiole entirely covering bud; buds with only one cap-like scale (Fig. 49 and p. 132).

  3. *With Yellow Patches*

   London Plane. Lvs. lobed; petiole entirely covering bud; buds with only one cap-like scale (Fig. 49 and p. 132).

  4. *With White Streaks*

   (1) Bladdernut. Lvs. with 3 lfts.; shrub with bladdery fr. (Pl. XXVII, 1, and p. 170).

   (2) Striped Maple. Lvs. lobed; bark green with white streaks; small tree (Pl. XXVIII and p. 175, no. 8).

  5. *With Curly Yellow Bark*

   Yellow Birch. Lvs. doubly serrate, simple, alternate; brts. with wintergreen taste (Pl. VIII, 2 and p. 89).

  6. *With Curly Cinnamon-Colored Bark*

   River Birch. Lvs. rhombic, glaucous beneath, alternate (Pl. VIII, 3 and p. 89).

  7. *Trunks Light Gray, Smooth*

   Beech. Lvs. simple, alternate; buds long-pointed (Pl. XII and pp. 96, 98).

  8. *Branches and Younger Parts of Trunk Light Gray*

   (1) Red Maple. Lvs. mainly 3-lobed, glaucous beneath, opposite (Pl. XXVIII, 7 and pp. 173, 174).

   (2) Siberian Elm. Lvs. simple, nearly simply serrate, alternate (Pl. XV, 4 and p. 114).

III. TRUNKS WITH WARTY GROWTHS

  Hackberry. Lvs. serrate, one-sided at base, alternate (Pl. XVI, 3 and p. 114).

IV. Brts. with Aromatic Odor
  1. Bayberry, Sweet Gale and Sweet-fern. Low shrubs; lvs., buds, and/or brts. with resin dots (Pl. VI, 1, 2, 3, and p. 83).
  2. Black and Yellow Birch. Brts. with wintergreen taste (Pl. VIII and p. 89).
  3. Magnolia. Large terminal buds; stipule scars encircling brt.; pith diaphragmed (Pl. XVIII and pp. 120–123).
  4. Mockernut Hickory. Large terminal bud; lvs. pinnate, usually 7 lfts., these and stout brts. pubescent (Pl. VII, 2 and p. 85).
  5. Prickly-ash. Lvs. pinnate, often prickly, with stipular thorns (Pl. XXIII, 2 and p. 159).
  6. *Prunus:* Cherry and Plum. Lvs. simple, petioles usually with glands (Pl. XX and pp. 146–150).
  7. Sassafras. Lvs. variously lobed or entire; brts. green (Pl. XIX, 2 and p. 126).
  8. Spicebush. Lvs. obovate; usually several small buds at one node (Pl. XIX, 1 and pp. 126, 127).
  9. Tulip-tree. Lvs. lobed, squarish at tip (Fig. 43, p. 124, and p. 123).
  10. Wafer-ash. Lvs. with 3 lfts. (Pl. XXIII, 3 and p. 159).
V. Brts. with Milky Sap (Seen best in the spring season)
  1. Mulberry Family: Osage-orange, with thorny brts. (Pl. XVII, 4, and p. 114); Mulberries (Pl. XVII, 1, 2, and p. 116); Paper-mulberry, lvs. lobed, sometimes opposite (Pl. XVII, 3 and pp. 114, 116).
  2. Norway and Hedge Maples. Lvs. palmately lobed, opposite (Pl. XXVIII and pp. 173, 174).
  3. Periwinkle (*Vinca*). Low, evergreen ground cover (p. 224).
  4. Sumacs: Staghorn and Smooth. Lvs. compound, alternate. (Dwarf S. and Poison-ivy sometimes slightly milky) (Pls. XXIV, XXV and pp. 161–165).
VI. Brts. Hollow (Lvs. opposite)
  1. Deutzia. (Fig. 47, p. 127).
  2. Forsythia spp. (Pl. XL, 1 and p. 221).
  3. Some Honeysuckles (Key, p. 240).
  4. Snowberry (p. 238).
VII. Brts. and Lvs. Covered with Silvery Scales
  Oleaster (Fig. 88, p. 186). See also other spp. of *Elaeagnus* and *Shepherdia.*
VIII. Brts. with Corky Ridges or "Wings"
  1. Rock or Cork Elm. Lvs. doubly serrate, alternate (Pl. XV, 3 and p. 114).
  2. Euonymus spp. Lvs. opposite (Pl. XXVII, 2, 4, and p. 169).
  3. Sweet-gum. Lvs. star-shaped, alternate (Pl. XIX, 3 and p. 129).
IX. Pith Colored (other than white)
  The pith of woody plants, by which I mean the soft tissue in the center of the stem, is normally white in the young stem. But in several spp. it takes on another color with age. To determine this character, then, we should cut across a two-yr.-old brt., using a sharp knife. This is sometimes difficult, as in tough brts. like those of hickory. The cut can however be easily made if the brt. is bent downward and the cut made from above at the center of the bend. In this way the lengthwise fibers of the stem are stretched and a sharp knife cuts through them easily. The following colors refer to the pith.

1. *Orange.* Red-berried Elder (Pl. XLIV, 2 and p. 231).
2. *Salmon-colored.* Coffee-tree (Pl. XXI, 2 and p. 153).
3. *Green.* Bayberry (Pl. VI, 1 and p. 83); Japanese Snowbell (Fig. 104, p. 215, and p. 215); Oleaster (Fig. 88, p. 186, and p. 186); Japanese Quince (Fig. 61, p. 140, and p. 141).
4. *White* with *green* border. Moonseed (Fig. 40 and pp. 119, 120).
5. *Brown:* Black Locust (Pl. XXII, 3 and p. 157); Sumacs (Staghorn and Smooth) (Pls. XXIV and XXV and pp. 161–165); Ailanthus (Pl. XXIV, 1 and p. 160); Smoke tree (Fig. 77 and p. 165); Grape (Fig. 83 and p. 177); Actinidia (Fig. 86 and p. 181); Silky Dogwood or Kinnikinnick, *Cornus Amomum* (Pl. XXXII, 1 and p. 193); Cork-tree (Pl. XXIII, 1, and p. 160); Asiatic Sweetleaf (p. 214 and fig. 103, a); Buttonbush (Pl. XLI, 2 and p. 230); Mt.-ash-spiraea, p. 134.

X. Pith Angled in Section
    1. Chestnut (Pl. XII and pp. 98, 99).
    2. Oak (Pls. XIII, XIV and pp. 99–110).

XI. Pith Chambered or with Diaphragms
    Lvs. opposite. Commonly cult. shrubs with simple, sometimes lobed or 3-divided lvs., and yellow fls. in early April (Pl. XL, 1 and p. 221).[*]
    Forsythia spp.
    Lvs. alternate.
    Vine. Actinidia (Fig. 86, p. 182, and p. 181).
    Trees or shrubs.
    Lvs. pinnately compound.
      Lfts. 11–17; pith dark-brown (Pl. VI, 4 and p. 83)........Butternut
      Lfts. 13–23; pith light-brown (Pl. VI, 5 and pp. 83, 85).Black Walnut
    Lvs. simple.
      Brts. aromatic; stipule scars encircling brts.
        Lvs. lobed (Fig. 43 and p. 123)...... ....Tulip-tree
        Lvs. not lobed; large, entire (Pl. XVIII, fig. 42, p. 123, and pp. 120–123)..................................Magnolia
      Brts. not as above.
        Lvs. entire, sometimes remotely toothed.
          Lvs. comparatively small (2½–5 in. long); large trees.
            Bark chunky; buds with 2 scales showing, one much overlapping; pith not always clearly chambered (Fig. 103, p. 213, and p. 212).....................Persimmon
            Bark not chunky; buds with many scales showing; pith with diaphragms at irregular intervals (Fig. 89, p. 187) Black-gum
          Lvs. larger (6–12 in. long); small trees. Terminal bud naked; pith sometimes indistinctly diaphragmed (p. 126)..Pawpaw
        Lvs. serrate.
          Bark warty; lvs. very one-sided, 2–5 in. long; usually many galls on lvs. and brts. (Pl. XVI, 3 and p. 114)..Hackberry
          Bark and lvs. not as above.
            Shrubs with small clusters of yellowish fls. in May; fr. an orange or brown drupe about ¼ in. long (p. 213).Sweetleaf

Trees.
> Lvs. 2–4 in. long; fls. white, bell-shaped, lining brts.
> (Fig. 105, p. 216, and p. 215).....Carolina Silverbell
> Lvs. 3–6 in. long; fls. peculiar, with two white, unequal
> bracts (Fig. 90, p. 188) .... Dove-tree

XII. Stems with Thorns, Prickles or Bristles

As explained at the beginning (p. 11), the word *thorn* is reserved for a modified organ—brt., lf., or stipules—while prickles and bristles are "emergences" appearing in no definite places or arrangement on the stem. Thorns are often not stable characters: sometimes they appear and sometimes they are absent, e. g. in the Black Locust, where however, they are usually conspicuous on young shoots; also in the Russian-olive or Oleaster, where they are often absent, etc. For our purpose this uncertainty is no drawback, however, since if no thorns are found on the plant there will be no occasion to use the following key, and the plant is included in the general keys (pp. 13 to 49). Prickles and bristles, however, are more constant characters.

A. Stems with thorns.
> 1. Thorns foliar, i. e. modified lvs. and therefore subtending brts. or buds.
> > Lvs. palmately compound (Pl. XXIX, 4, and p. 188).
> > > Acanthopanax
> > Lvs. simple (Fig. 39, p. 119)......................Barberry
> 2. Thorns stipular, i. e. modified stipules, therefore in pairs at lf. base.
> > Lvs. abruptly pinnate; fls. yellow, of pea type (p. 159)..Pea-tree
> > Lvs. odd pinnate.
> > > Lfts. pointed but not mucronate (Pl. XXIII, 2 and p. 159).
> > > Prickly-Ash
> > > Lfts. rounded, mucronate (Pl. XXII, 3, and pp. 157, and 159).
> > > Black and Clammy Locusts
> 3. Thorns modified brts. and therefore in axils of lvs., subtended by lf. scars, or at the tip of a brt.
> > Thorns in axils of lvs., subtended by lf. scar, or on trunk.
> > > Lvs. simple, thorns unbranched.
> > > > Lvs. entire; thorns short (Fig. 110, p. 225 and p. 227).
> > > > Matrimony Vine
> > > > Lvs. serrate.
> > > > > Lvs. evergreen; thorns short (Fig. 54, p. 136, and p. 135).
> > > > > Firethorn
> > > > > Lvs. deciduous, often lobed; thorns larger (p. 135).
> > > > > Hawthorn
> > > Lvs. either once or twice pinnate; thorns usually branched
> > > (Fig. 70, p. 152, and p. 153)............Honey-locust
> > Thorns at tip of brt. which may or may not be leafy.
> > > Stems covered with silvery scales (Fig. 88, p. 186).
> > > > Oleaster
> > > Stems not as above.
> > > > Lvs. entire, on long arching brts.; vine-like shrub with purple fls., sometimes escaped from cult. (p. 227).
> > > > Matrimony Vine

Lvs. serrate.

Stipules prominent and persistent; ornamental shrub with greenish fr. (Fig. 61 p. 140, and p. 141).

Japanese Flowering-quince

Stipules not as above.

Fr. a pome; brt. often leafy, ending in strong thorn (Fig. 62, p. 141)......................Pear

Fr. berry-like, black (Fig. 82, p. 177, and p. 176).

Buckthorn

B. Stems with prickles.

1. Lvs. simple (Fig. 24 and p. 73)..Greenbrier and other Smilax spp.

2. Lvs. simple, lobed; prickles often larger under lf. scars (Fig. 48, p. 128.......................................Gooseberries

3. Lvs. with 3-5 lfts.; stems and often lvs. prickly. Blackberries (p. 144)

4. Lvs. once pinnate; stems and sometimes lvs. prickly (Fig. 67, p. 146, and pp. 145, 146)...........................Roses

5. Lvs. very large, twice pinnate; trunk, brts., and lvs. prickly (Fig. 92, p. 192, and p. 189).......................Hercules-club

C. Stems with bristles.

1. Lvs. simple (Fig. 24 and p. 73)..Sawbrier and other Smilax spp.

2. Lvs. with usually 3 to 5 lfts.; stems and lvs. bristly (Fig. 65, p. 144, and p. 143)......................Raspberry and Wineberry

3. Lvs. pinnately compound (Pl. XXII, 1 and p. 159)..Bristly Locust

XIII. CLIMBING STEMS (VINES)

1. Climbing by twining stems (sometimes merely leaning as in Bitter Nightshade or Climbing Roses).

Lvs. simple, opposite.

Fls. yellowish white in dense clusters at or near ends of brts. (p. 240).

Woodbine Honeysuckle

Fls. red (Pl. XLV, 3 and p. 240)...........Trumpet Honeysuckle

Fls. white, turning yellow (Pl. XLV, 4 and p. 240).

Japanese Honeysuckle

Lvs. simple, alternate, not lobed.

Pith chambered; fr. like that of grape, but green, edible; rarely cult. vine (Fig. 86, p. 182, and p. 181).............Actinidia

Pith not chambered.

Lvs. very large, kidney-shaped, with entire margins; fls. pipe-shaped (Fig. 36, p. 117)...............Dutchman's-pipe

Lvs. ovate to oblong-ovate, or more or less rounded; frs. yellow-skinned, with crimson interior (Pl. XXVII, 3 and p. 170).

Climbing Bittersweet, both native and oriental spp.

Lvs. simple, alternate, with whole margin more or less lobed.

Rather rare, native vine in rich woods; lf. scar raised and concave (Fig. 40, p. 120, and p. 119)....................Moonseed

Lvs. often pinnately lobed at base; berries red, poisonous (Pl. XLI, 1 and p. 225)..............................Bitter Nightshade

Lvs. lobed at base, as in an arrowhead (hastate), margins wavy; fls. and fr. abundant, white or greenish in late summer or early fall; sometimes cult. (p. 117)...............................Fleecevine

Lvs. pinnately compound, alternate.
   Stems prickly (p. 146)..........................Climbing Roses
   Stems not prickly; fls. of pea type, blue or white (Fig. 73, p. 157)
                                         Wisteria
Lvs. palmately compound, alternate, with 5 lfts., rarely cult. (Fig. 38
   and p. 118)...Akebia
2. Climbing by aerial rootlets. Cf. figs. 2, p. 7, and 111, p. 229, and Pl.
   XXV, 2.
   Lvs. opposite.
      Lvs. simple, evergreen; fls. inconspicuous, greenish (p. 170).
                             Winter-creeper Euonymus
      Lvs. simple, deciduous; fls. small, white, in clusters, with some fls.
         enlarged and sterile (p. 128)...........Climbing Hydrangea
      Lvs. pinnate, deciduous; fls. large, tubular, orange (Fig. 111, p.
        229, and p. 227)........................Trumpet-creeper
   Lvs. alternate.
      Lvs. simple, lobed, evergreen (Fig. 91, p. 189, and p. 188)
                                 English Ivy
      Lvs. compound with 3 lfts., deciduous, poisonous to touch (Pl.
        XXV, 2 and p. 165).........................Poison-ivy
3. Climbing by tendrils.
   Tendrils stipular and simple, i. e. modified stipules, a pair at the base
   of each lf.; stems usually prickly or bristly (Fig. 24, p. 73).
                                    Smilax
   Tendrils not stipular.
      Lvs. simple, alternate, sometimes lobed.
        Pith brown (Fig. 83, p. 179, and p. 177)..............Grape
        Pith white; lvs. 3-lobed.
          Tendrils without disk-like tips (p. 177).
                     Porcelain-berry (*Ampelopsis*)
          Tendrils with disk-like, adhesive tips (Pl. XXX, 1 and
           p. 179)................................Boston-ivy
      Lvs. compound.
        Lfts. 3 (Pl. XXX, 1 and p. 179)................Boston-ivy
        Lfts. usually 5; tendrils usually with disk-like tips; fr. blue-
        black, with a bloom; native (Pl. XXX, 2 and pp. 177–179).
                             Virginia Creeper
4. Climbing by twining lf. stalks (petioles) (p. 118).............Clematis

SOME HELPFUL REFERENCES

1. AMERICAN JOINT COMMITTEE ON HORTICULTURAL NOMENCLATURE. Standardized Plant Names. XVI + 675 pp. 2nd ed. J. Horace McFarland Co.: Harrisburg, Pa., 1942.
2. BAILEY, L. H. The cultivated conifers in North America. XII + 404 pp. illus. The Macmillan Co.: New York, 1933.
3. ————. The standard cyclopedia of horticulture. 3 vols. illus. The Macmillan Co.: New York, 1943.
4. BLACKBURN, BENJAMIN. Trees and shrubs in eastern North America. Keys to native and cultivated woody plants growing in the temperate regions, exclusive of conifers. Oxford Univ. Press: New York, 1952.
5. BLAKESLEE, A. F., AND C. D. JARVIS. Trees in winter, their study, planting, care, and identification. 446 pp. illus. The Macmillan Co.: New York, 1913.
6. ————. Trees in winter. 292 pp., 516 figs. The Macmillan Co.: New York, 1931.
7. BRITTON, N. L. Manual of the flora of the northern States and Canada, 3rd ed. XXIV + 1122 pp. Henry Holt & Co.: New York, 1907.
8. BROWN, H. P. Trees of northeastern United States, native and naturalized. Rev. ed. 490 pp. illus. Christopher Publ. House: Boston, 1938.
9. BURNS, GEORGE P., AND CHARLES H. OTIS. The trees of Vermont. Vermont Agric. Exp. Sta. Bull. 194. 244 pp. illus. Free Press Printing Co.: Burlington, Vt., 1916.
10. COLLINGWOOD, G. H. Knowing your trees. 213 pp. illus. American Forestry Association: Washington, D.C., 1947.
11. COMMITTEE OF THE VERMONT BOTANICAL CLUB, E. J. Dole, Editor. The flora of Vermont. 3rd rev. ed. 553 pp. Free Press Printing Co.: Burlington, Vt., 1937.
12. DAME, LORIN L., AND HENRY BROOKS. Handbook of the trees of New England. 196 pp. illus. Ginn and Co., The Athenaeum Press: Boston, 1902.
13. ENGLER, A., AND K. PRANTL. Die natürlichen Pflanzenfamilien. Part 2. Section 1. Coniferae by A. W. Eichler. pp. 28–116. W. Engelmann: Leipzig, 1889.
14. FERNALD, MERRITT LYNDON. Gray's new manual of botany. 8th ed. LXIV + 1632 pp. illus. American Book Co.: New York, 1950.
15. FOSTER, JOHN H. Trees and shrubs of New Hampshire. 2nd ed. illus. 117 pp. Published by Society for the Protection of New Hamsphire Forests: Concord, N. H., 1941.
16. GLEASON, HENRY ALLAN. The new Britton and Brown illustrated flora of the northeastern United States and adjacent Canada. 3 vols. New York Botanical Garden: 1952.
17. GRAVES, ARTHUR HARMOUNT. Woody plants of Brooklin, Maine. Rhodora 12: 173–184. 1910.
18. ————. Various illustrated papers written for the School Nature League, N. Y. City and now published by the National Audubon Society, 1000 Fifth Ave., N. Y. as follows:

    (1). Common oak trees in winter. Ser. 7, no. 5. 1937.
    (2). Street trees. Ser. 8, no. 9, 1938. 4th printing: 1951.
    (3). Maples of N. Y. and vicinity. Ser. 9, no. 9. 1939.
    (4). Common oak trees in summer. Ser. 10, no. 10. 1940.

(5). Birches of the N. Y. City region. Ser. 11, no. 10. 1941.
(6). Pines of the N. Y. City region. Ser. 12, no. 6. 1942.
(7). Spruces of the eastern U. S. Ser. 13, no. 6. 1943.

19. ————. Winter key to woody plants of the northeastern United States and adjacent Canada. 33 pp. Publ. by author, Wallingford, Conn., 1955.

20. ———— AND HESTER M. RUSK. Guide to trees and shrubs, based on those of Greater New York; native, naturalized and commonly cultivated exotic kinds. XII + 76 pp. 4th printing, 1949. Publ. by Hester M. Rusk, Brooklyn Botanic Garden, Brooklyn, N. Y.

21. GRAVES, CHARLES BURR, AND OTHERS. Catalog of the flowering plants and ferns of Connecticut growing without cultivation. 569 pp. Bull. 14. State Geological and Natural History Survey, State Library, Hartford, Conn., 1910. Also Bull. 48, Additions to the flora of Conn. by E. B. Harger and others, a supplement to Bull 14. 94 pp. Publ. by State of Conn., Hartford, Conn., 1930.

22. Graves, G. Trees, shrubs, and vines for the northeastern United States. Oxford Univ. Press: New York, 1945.

23. GRIMM, WILLIAM C. The shrubs of Pennsylvania. 522 pp. Stackpole: Harrisburg, Pa., 1952.

24. ————. Trees of Pennsylvania. 363 pp. illus. Stackpole and Heck, Telegram Press Bldg.: Harrisburg, Pa., 1950.

25. HARLOW, W. M. Twig key to the deciduous woody plants of Eastern North America. 4th ed. Publ. by author: Syracuse, N. Y., 1948.

26. ———— AND ELLWOOD S. HARRAR. Text book of dendrology. 3rd ed. XIII + 555 pp. McGraw-Hill Book Co.: New York, 1950.

27. HOUGH, R. B. Handbook of the trees of the northern States and Canada east of the Rocky mountains. X + 470 pp., 498 figs. Publ. by the author: Lowville, N. Y., 1924.

28. HOUSE, HOMER D. Annotated list of the ferns and flowering plants of New York State. N. Y. State Museum Bull. 254. 759 pp. Univ. of State of N. Y.: Albany, N. Y., 1924.

29. HURFORD, A. W. Forest trees common to southern New England and adjacent areas in New York State (revision of Forest trees of Connecticut by Austin F. Hawes and Wilbur R. Mattoon). Conn. Forest and Park Association; New Haven, Conn., 1955.

30. HYLAND, FAY. The conifers of Maine. Maine extension Bull. 345. 20 pp. Agric. Extension Service, Univ. of Maine and U.S.D.A.: Orono, Maine. Repr. 1949.

31. ———— AND FERDINAND H. STEINMETZ. The woody plants of Maine, their occurrence and distribution. Univ. of Maine Studies, Second Ser., No. 59. IX + 65 pp. Univ. Press: Orono, Maine, 1944.

32. ILLICK, JOSEPH S. Common trees of New York. 123 pp. illus. American Tree Association: Washington, D. C., 1927.

33. ————. Tree habits: how to know the hardwoods. 341 pp. American Tree Association: Washington, D. C., 1924.

34. KEELER, HARRIET L. Our native trees and how to identify them. XXIII + 533 pp. illus. Chas. Scribner's Sons: New York, 1912.

35. ————. Our northern shrubs and how to identify them. XXX + 521 pp. illus. Chas. Scribner's Sons: New York, 1928.

36. LAWRENCE, GEORGE H. M. Taxonomy of vascular plants. IX + 823 pp. The Macmillan Co.: New York, 1951.

37. LITTLE, ELBERT L., JR. Check list of native and naturalized trees of the United States (including Alaska). Agriculture Handbook No. 41. 472 pp. U. S. Government Printing Office: Washington, D. C., 1953.

38. MAKINS, F. K. The identification of trees and shrubs. VII + 326 pp. illus. E. P. Dutton & Co.: New York, 1937.

39. MOLDENKE, H. N. American Wild Flowers. XXV + 453 pp. illus. D. Van Nostrand Co. Inc.: New York, 1950.

40. MUENSCHER, W. C. Keys to woody plants. 5th ed. 108 pp. Comstock Publishing Co.: Ithaca, N. Y. 1946.

41. OGDEN, E. C., F. H. STEINMETZ, AND F. HYLAND. Check-list of the vascular plants of Maine. Bull. of the Josselyn Botanical Society, and available through its President. Photographed by Spaulding Moss Co.: Boston, Mass. 1948.

42. OTIS, CHARLES H., AND GEO. P. BURNS. Michigan trees. A handbook of the native and most important introduced species. 362 pp. illus. Publ. by Regents of Univ. of Michigan: Ann Arbor. 1931.

43. PEATTIE, DONALD CULROSS. A natural history of trees of Eastern and Central North America. XV + 606 pp. illus. Houghton Mifflin & Co.: Boston. 1950.

44. PERKINS, HAROLD O. Street trees for Connecticut. Folder 68. Agric. Extension Service of Univ. of Connecticut. Storrs, Conn., 1953.

45. ———. Trees, shrubs and vines for Connecticut. Agric. Extension Service. College of Agric. Univ. of Connecticut. Bull. No. 55–17. March, 1955.

46. PLATT, RUTHERFORD. American trees. 256 pp. illus. Dodd, Mead: New York. 1952.

47. PORTER, THOMAS CONRAD. Flora of Pennsylvania, edited with the addition of analytical keys by John Kunkel Small. XV + 362 pp. Ginn & Co.: Boston, 1913.

48. PRESTON, RICHARD J., JR. North American trees. IV + 371 pp. illus. Iowa State College Press: Ames, Iowa. 1948.

49. REHDER, ALFRED. Manual of cultivated trees and shrubs. 2nd ed. XXX + 996 pp. The Macmillan Co.: New York. 1940.

50. ROBINSON, FLORENCE BELL. Tabular keys for the identification of the woody plants. The Garrard Press: Champaign, Ill., 1941.

51. ROGERS, JULIA E. The tree book, a popular guide to a knowledge of the trees of North America and to their uses and cultivation. 565 pp. illus., Doubleday, Doran & Co.: New York, 1931.

52. ROGERS, MATILDA. A first book of tree identification. 95 pp. illus. Random House: New York, 1951.

53. SARGENT, CHARLES S. Manual of the trees of North America exclusive of Mexico. 2nd ed. VIII + 910 pp. 783 figs. Houghton Mifflin Co.: Boston and New York. 1922.

54. ———. The silva of North America. 14 vols. 4to. Houghton Mifflin Co.: Boston and New York. 1894–1902.

55. SCHAFFNER, JOHN H. Field manual of trees. 4th ed. 160 pp. R. G. Adams and Co.: Columbus, Ohio, 1936.

56. SCHNEIDER, CAMILLO KARL. Dendrologische Winterstudien. VI + 290 pp. illus. Gustav Fischer; Jena, 1903. Out of print, but available in large botanical or forestry libraries.

57. SINNOTT, EDMUND W. Botany: principles and problems. 5th ed. X + 528 pp. McGraw-Hill Book Co. Inc.: New York, 1955.

58. STONE, WITMER. The plants of southern New Jersey with especial reference to the flora of the Pine Barrens and the geographic distribution of the species. 809 pp. and 129 pls. Trenton, N. J., 1911.

59. TRELEASE, WILLIAM. Plant materials of decorative gardening. 204 pp. Publ. by the author: Urbana, Ill., 1917.

60. ————. Winter botany. XI + 396 pp. illus. Publ. by the author: Urbana, Ill., 1917.

61. WYMAN, DONALD. Shrubs and vines for American gardens. IV + 442 pp. illus. The Macmillan Co.: New York, 1949.

62. YEARBOOK COMMITTEE. U. S. Dept. of Agric. Trees, the yearbook of agriculture, 1949. A general account of trees, their uses, and of the forests. XIV + 994 pp. Govt. Printing Office: Washington, D. C.

GLOSSARY

Accessory buds. Buds near the nodes but not in the leaf axils.

Achene. A small, dry, one-seeded, indehiscent fruit.

Acicular. Needle-shaped.

Acute. Sharp-pointed.

Alternate. Not opposite; (leaves) only one at a node; (buds) only a single *axillary* bud at a node.

Ament. Catkin.

Aril. A fleshy outgrowth from a seed, sometimes more or less surrounding it.

Awl-shaped. Tapering from a thick base to a sharp point.

Axil. Angle; e. g. the angle between a leaf and stem, or the angle of a branching vein.

Axillary. Situated in an axil.

Axis. The center line of any organ, or the central organ around which others are attached.

Berry. A many-seeded fleshy fruit.

Bipinnate. Twice pinnate. See under pinnate.

Blade. The expanded part of a leaf.

Bract. A small leaf or scale, in the axil of which a flower, flower cluster, or cone scale may be borne.

Branch. A subdivision of the main stem.

Branchlet. The growth of the last season on any stem.

Bristle. A stiff hair.

Bud. An undeveloped stem with undeveloped leaves, or flowers, or both.

Bud scales. Small, dry, modified leaves covering a bud.

Bundle scars. Small marks on a leaf scar where the vascular bundles (conducting strands) passed from the stem and connected with the veins in the leaf.

Calyx. A collective name for the sepals.

Capsule. A dry, dehiscent fruit developed from a compound ovary.

Catkin. A spike of unisexual flowers, each borne in the axil of a bract.

Chambered pith. With cavities separated by plates or disks, distinct from "diaphragmed pith" where there are no cavities.

Ciliate. Fringed with hairs.

Close bark. Not scaly.

Collateral buds. Accessory buds at the sides of the axillary bud.

Compound leaf. A leaf whose blade is divided into separate parts called leaflets.

Cone. A spike-like cluster of scales bearing naked seeds.

Conifers. Cone-bearing trees of the Pine family.

Connate. Grown together.

Cordate. (Leaves) heart-shaped at the base.

Corolla. Collective name for the petals.

Crenate. Scalloped, or with rounded teeth.

Crenulate. Slightly crenate.

Cuneate. Wedge-shaped at base.

Deciduous. Falling off; applied to leaves which drop off in the autumn of their first year.

Decurrent. (Leaf) extending down the stem below the place of insertion.

Decussate. Arranged in pairs which alternate with each other at right angles, making four vertical ranks.

Dehiscent. Splitting open when ripe.

Diaphragmed pith. With transverse firmer plates of tissue.

Dioecious. Having staminate and pistillate flowers on separate plants.

Distichous. Alternate, in 2 ranks.

Downy. Covered with short, soft hairs.

Drupe. A stone fruit, the fleshy part surrounding a stone which encloses the seed, as in the peach: typically with one stone but sometimes with more than one.

Elliptic. About twice as long as wide, and with the general outline of an ellipse, the two ends about the same width.

Emarginate. With a shallow notch at the tip.

Entire. With an even margin, not toothed, or divided.

Established. Reproducing and growing of itself without cultivation.

Evergreen. (Leaves) remaining green through the winter.

Excurrent. With reference to tree growth; having a single central trunk.

Exfoliating. Peeling off in layers.

Exotic. Of foreign origin and not naturalized.

Falcate. Sickle- or scythe-shaped.

Fascicle. A bundle or close cluster.

Fimbriate. Fringed.

Fluted. With rounded ridges.

Fruit. A ripened ovary or seed vessel.

Glabrate. Formerly pubescent, now glabrous.

Glabrous. Smooth, i. e. without hairs.

Gland. A secreting organ, embedded, or mounted on a stalk, or tipping a hair or tooth; or any protuberance resembling such an organ.

Glandular. Furnished with glands.

Glaucous. Covered with white or bluish bloom which rubs off, as on a plum.

Globose. Spherical or nearly so.

Glutinous. With a sticky exudation.

Hairy. With fairly long hairs.

Half-evergreen. (Leaves) remaining green through part of the winter.

Imbricated. Overlapping.

Indehiscent. Not splitting open when ripe.

Internode. The part of the stem between two nodes.

Involucre. A group of modified leaves around a flower or flower cluster; and so, around the fruit.

Keeled. With a central ridge, like the keel of a boat.

Lanceolate. Shaped like the head of a lance; much longer than wide, tapering to a point at the upper end, and slightly narrowed at the base.

Lateral. Situated on the side.

Leader. The trunk of a tree.

Leaf bud. A bud containing leaves and stem, but no flowers.

Leaf scar. A scar left where a leaf fell from the stem.

Leaflet. One of the parts of a compound leaf.

Legume. A dry fruit formed from a simple ovary, and splitting into 2 valves when ripe, as the pea pod.

Lenticel. A raised dot marking a region of loose aerating tissue in the bark.

Linear. Long and very narrow, with parallel margins.

Lobe. A segment of a leaf whose margin is too deeply cut to be called toothed.

Long growth. That part of the brt. which extends it for the year; as opposed to the short, side, spur-like growths. See pp. 11 and 12.

Metamorphosed. Changed in form.

Midrib. The central vein of a leaf.

Mixed bud. A bud containing both leaves and flowers.

Monoecious. Having staminate and pistillate flowers on the same plant.

Mucronate. Abruptly tipped with a short, blunt point. *Mucronulate*, with a very short point.

Multiple buds. Many extra buds at one node.

Multiple fruit. A structure composed of the fruits of several flowers so close together as to appear as a single fruit.

Naked. Not covered: (bud) without specially modified, covering scales; (catkin) not enclosed in a bud; (seed) not enclosed in an ovary.

Naturalized. Growing and reproducing without cultivation and originally from a foreign area.

Needle. A long, slender, more or less needle-shaped leaf.

Node. A place on the stem where one or more leaves are (or were) borne.

Nut. A hard, mostly one-seeded, indehiscent fruit, larger than an achene.

Nutlet. A very small, nut-like fruit.

Oblanceolate. Lanceolate with the tapering point toward the base.

Oblong. Longer than wide, with margins nearly parallel.

Obovate. Ovate with the narrow end toward the base.

Obtuse. Blunt or rounded at the tip.

Opposite. (Leaves or axillary buds) two at a node, inserted on opposite sides of the stem at the same level.

Oval. Broadly elliptical, less than twice as long as wide.

Ovary. The part of the flower in which the seeds will be formed; *compound ovary*, one composed of 2 or more parts.

Ovate. With the general outline of a lengthwise section of an egg, with the wider end toward the base.

Ovulate. With ovules.

Ovule. Young, usually unfertilized seed.

Palmate. Resembling a hand: *palmately* veined, with three or more veins about the same size arising from the same point at the base of the blade; *-lobed*, with sinuses pointing toward the petiole; *-compound*, with leaflets all attached to the tip of the petiole.

Panicle. An elongated, loose, branching or compound flower cluster.

Papilionaceous. With flowers constructed like those of the sweet pea.

Pappus. The modified calyx in the Composite family.

Parallel veined. With veins nearly parallel.

Parasitic. Getting food partly or wholly from another living organism.

Parted. (2-, 3- etc. parted): too deeply cut to be called lobed, but not deeply enough to be called compound.

Peltate. Shield-shaped, attached to stalk inside the margin.

Pendulous. Drooping.

Perennial. Lasting year after year.

Persistent. Remaining on; not deciduous; applied to leaves, pubescence, etc.

Petal. One of the modified leaves (usually bright-colored) forming the inner circle of leaf-like parts of a flower, next to and surrounding the stamens.

Petiole. The stalk of a leaf.

Phyllotaxy. The arrangement of leaves on the stem.

Pinnate. Resembling a feather; *pinnately* veined, lobed, or compound, with veins, lobes, or leaflets arranged along the sides of a central axis or rachis. *Twice pinnate.* where the leaflets themselves are also pinnately compound. *Odd pinnate.* With a terminal leaflet. *Abruptly pinnate.* Ending with a pair of leaflets, no terminal leaflet.

Pistil. The central organ of a flower, in the base (ovary) of which the seeds will be formed.

Pistillate. Having one or more pistils, but no stamens.

Pith. The softer, central part of a stem.

Plumose. Feathery.

Pollen. A powder, usually yellow, discharged from the enlarged tips (anthers) of the stamens of a flower.

Pome. A fleshy fruit on the plan of an apple or pear.

Prickle. A slender, sharp-pointed outgrowth from the young bark or epidermis, an emergence.

Puberulous, or *puberulent.* Minutely pubescent.

Pubescent. Bearing hairs of some sort, soft and fine ones particularly.

Raceme. A cluster of stalked flowers on an elongated axis.

Racemose. Raceme-like.

Rachis. The axis of a compound leaf (or of a spike).

Receptacle. The tip of the stem (usually somewhat enlarged) on which the parts of the flower are borne.

Resin duct. A lengthwise or transverse canal which carries resin.

Rhombic. With 4 nearly equal sides, but not rectangular.

Rootstock. An underground stem, often thickened. A rhizome.

Rugose. Wrinkled.

Samara. A winged fruit.

Scabrous. Rough to the touch.

Scale. (1) A very small leaf, usually appressed and often dry; e.g., modified leaves that cover buds; modified leaves of cones and catkins.

    (2) A tiny flattened outgrowth from the epidermis.

    (3) A flake of bark.

Serrate. Saw-toothed; with sharp teeth pointing forward.

Serrulate. Finely toothed.

Sessile. Without a stalk.

Shoot. Stem and leaves.

Shrub. A woody plant, branched from the base, usually less than 15 ft. tall.

Simple. Not branched; not compound.

Sinus. The indentation between two lobes.

Spike. A cluster of sessile flowers borne close together on an elongated axis.

Stamen. One of the pollen-bearing organs of a flower.

Staminate. (Fls.) Having stamens but no pistils.

Stellate pubescent. With star-like hairs, i.e. hairs originating from a single point.

Stem. The trunk, branch, branchlet or twig of the plant; not the stalk of the leaf, which is the petiole.

Stipel. Appendage to leaflet analogous to stipule.

Stipules. Small appendages occurring in pairs at the bases of the petioles of the leaves of certain plants.

Stipule scar. A scar left on the stem where a stipule had been.

Striate. Marked with lengthwise stripes or ridges.

Strigose. With appressed bristles.

Style. The usually slender upper part of the pistil connecting stigma and ovary.

Subopposite. Not quite opposite.

Subpetiolar. Underneath the base of the petiole.

Sub-shrub. A low plant which is somewhat or slightly woody.

Subtend. Inserted just below.

Superposed buds. Arranged one above the other(s) above the leaf axil.

Tendril. A thread-like organ (modified stem or leaf or part of leaf) which coils around or is fastened to a support.

Terete. Cylindrical.

Thorn. A fairly long, slender, hard, sharp-pointed structure, in this book a modified organ—stem, leaf or stipule.

Tomentose. Covered with a woolly felt (tomentum).

Tomentulose. Finely tomentose.

Toothed. With short projections between shallow notches on the margin.

Tree. A woody plant usually with one main trunk, and reaching a height of at least 15 feet.

Trunk. The main stem of a tree.

Twig. A small branch, usually including several years' growth.

Umbel. Umbrella-like flower cluster; with flower stalks all from one point, the tip of the axis.

Undulate. With a wavy surface or margin

Unisexual. Staminate or pistillate.

Valvate. With edges meeting and not overlapping.

Valve. One of the pieces into which a dehiscent fruit splits.

Variegated. With marks or patches of different colors.

Veins. Strands of conducting tissue forming the framework of leaves.

Villous. With long, soft hairs.

Whorl. A circle of three or more (leaves or buds) around the stem.

Wing. A thin expansion of, or appendage to, an organ.

Woolly. Covered with long, entangled, soft hairs.

INDEX

Descriptions and illustrations are on pages 50 to 242, general keys on pages 13 to 49. Pages on which illustrations occur are in **b o l d face** type; synonyms and other botanical names not used by the author are in *italics*. Incidental references scattered through the text, and names in the keys, are not included, except in the Honeysuckle genus.